Microbial Ecological Theory

Current Perspectives

Edited by

Lesley A. Ogilvie

University of Brighton
Brighton
UK

and

Penny R. Hirsch

Rothamsted Research
Harpenden
UK

Caister Academic Press

Caister Academic Press
Norfolk, UK

www.caister.com

British Library Cataloguing-in-Publication Data
A catalogue record for this book is available from the British Library

ISBN: 978-1-908230-09-6

Description or mention of instrumentation, software, or other products in this book does not imply endorsement by the author or publisher. The author and publisher do not assume responsibility for the validity of any products or procedures mentioned or described in this book or for the consequences of their use.

Cover image courtesy of Martin Riese.

Printed and bound in Great Britain

Contents

Contributors

Zaid Abdo
Institute of Bioinformatics and Evolutionary Studies;
Department of Mathematics and Statistics
University of Idaho
Moscow, ID
USA

zabdo@uidaho.edu

Diego Fontaneto
National Research Council
Institute of Ecosystem Study
Verbania Pallanza
Italy

d.fontaneto@ise.cnr.it

Larry J. Forney
Institute of Bioinformatics and Evolutionary Studies;
Department of Biological Sciences
University of Idaho
Moscow, ID
USA

lforney@uidaho.edu

Christopher J. van der Gast
NERC Centre for Ecology and Hydrology
Wallingford
UK

cjvdg@ceh.ac.uk

Jiawei Geng
The First People's Hospital of Yunnan Province
Kunming
People's Republic of China

sgbox2002@yahoo.com.cn

Penny R. Hirsch
Rothamsted Research
Harpenden
UK

penny.hirsch@rothamsted.ac.uk

Joaquín Hortal
Departamento Biodiversidad y Biología Evolutiva
Museo Nacional de Ciencias Naturales (CSIC)
Madrid
Spain;
Depto. Ecologia
Universidade Federal de Goiás
Gôiania
Brazil

jhortal@mncn.csic.es

Brian V. Jones
Centre for Biomedical and Health Science Research
School of Pharmacy and Biomolecular Sciences
University of Brighton
Brighton
UK

b.v.jones@brighton.ac.uk

Andrew K. Lilley
Molecular Microbiology Research Laboratory
Pharmaceutical Science Division
King's College London
London
UK

andy.lilley@kcl.ac.uk

Zhanshan (Sam) Ma
Computational Biology and Medical Ecology Lab
State Key Laboratory of Genetic Resources and Evolution
Kumming Institute of Zoology
Chinese Academy of Sciences
Kumming
People's Republic of China;
Institute of Bioinformatics and Evolutionary Studies
Department of Biological Sciences
University of Idaho
Moscow, ID
USA

ma@vandals.uidaho.edu

Tim H. Mauchline
Rothamsted Research
Harpenden
UK

tim.mauchline@rothamsted.ac.uk

Lesley A. Ogilvie
Centre for Biomedical and Health Science Research
School of Pharmacy and Biomolecular Sciences
University of Brighton
Brighton
UK

l.ogilvie@brighton.ac.uk

Anna Oliver
NERC Centre for Ecology and Hydrology
Wallingford
UK

aela@ceh.ac.uk

Andrew D.J. Overall
Centre for Biomedical and Health Science Research
School of Pharmacy and Biomolecular Sciences
University of Brighton
Brighton
UK

a.d.j.overall@brighton.ac.uk

Paul Wilmes
Luxembourg Centre for Systems Biomedicine
University of Luxembourg
Esch-sur-Alzette
Luxembourg

paul.wilmes@uni.lu

Preface

The vast explosion of high-resolution molecular data in the past few years has provided an unprecedented glimpse into the microbial world – with tantalizing results. Thus, the time is right to delve deeper into the ever-increasing knowledge base on this unseen majority. Born out of a desire to provide further insight into this accumulating wealth of data, the book synthesizes current view points and knowledge on the topic of microbial ecological theory. We have assembled a collection of essays by a diverse group of well-respected scientists who merge the boundaries of ecology and microbiology, to explore some of the central tenets of macro-ecological theory with a microbial perspective. The contributors explore the mainstays of macro-ecology asking questions such as 'Are microbes everywhere?' and 'Does a microbial species concept exist?', as well as showing how high-resolution molecular data are informing and underpinning the evolution of microbial ecological theory. What becomes apparent is how the application of macro-ecological theory to the microbial world is not only enhancing our understanding of microbial ecology but also providing a reference point for the development of new theories.

Written for graduate students and academic researchers, the book aims to stir cross-disciplinary thinking and provide direction and perspective on the still fledgling field of microbial ecological theory.

Lesley A Ogilvie is a Senior Research Fellow at the University of Brighton, UK. Her research is focused on microbial ecology, specifically the connection between gut microbial community structure and functioning and the onset and progression of disease.

Penny R Hirsch leads a research group at Rothamsted Research, supported by BBSRC funding. Her interests include the microbial ecology of soils and plants, in particular those related to improving the sustainability of food production.

The editors would like to wholeheartedly thank all authors for their contributions to this book, acknowledging the time, effort , and original thought put into each chapter.

Lesley A. Ogilvie and Penny R. Hirsch

Genome-based and Functional Differentiation: Hallmarks of Microbial Adaptation, Divergence and Speciation?

1

Paul Wilmes

Abstract

The recent application of high-throughput molecular biology methods to natural microbial communities is profoundly changing our view on the microbial world. In particular, our understanding of microbial population-level differentiation involved in ecological adaptation that leads to microbial divergence and speciation has been profoundly altered. Numerous processes that underlie microbial differentiation have been identified but determining the relative significance of these processes remains challenging. For example, a major unresolved question is how much of observed genetic heterogeneity is due to neutral versus adaptive processes. Sequence-based and modelling analyses suggest that much of the observed variation is neutral but recent functional 'meta-OMIC' data suggest that at least some of this fine-scale variation is functionally relevant and, thus, involved in adaption and divergence. From the limited amount of largely disjointed metagenomic and functional data obtained to date, extensive intra- and inter-system as well as extensive intra- and inter-population differences are apparent. Consequently, it is difficult to ascertain generalizable rules that delineate specific microbial groups that would be congruent with the definition of a microbial species. Future concomitant analysis of community genomic complements, transcriptomes, proteomes and metabolomes over relevant spatial and temporal scales will result in detailed molecular descriptions of distinct taxonomic entities. Such a system-level molecular organismal classification system will need to be solidly grounded in ecological theory, population genetic theory and evolutionary theory, and may be universally applicable to the three domains of life.

Introduction

Since early observations by Antonie van Leeuwenhoek, we know that in nature microorganisms form multi-organism assemblages that represent ecological communities. However, the vast majority of our knowledge to date concerning microbial biochemistry and genetics, which drive microbial evolution and define their ecology, has been largely based on single isolated strains. These are for the most part obtained by harsh selective pressures in the laboratory environment and, thus, are often not representative of their natural population context. In the natural environment, microbial populations comprise individuals that are distinct from each other but can be classified into congruent groups based on several criteria (Achtman and Wagner, 2008). The application of high-resolution molecular biology techniques directly to environmental samples, ranging from deep small subunit ribosomal gene sequencing to most recent 'meta-OMICS' efforts, is allowing unprecedented surveys of natural microbial population structure and function. The results of such efforts are inviting us to revise or adapt current paradigms in biology.

A key question that has occupied microbiologists and evolutionary biologists is if microbial diversification and evolution can be framed in a concept similar to that used to describe eukaryote systematics (Cohan, 2002). Since the middle of the nineteenth century, microbiologists have been using binominal Linnaean names to designate microbial species by assuming that this classification offers enough resolution to distinguish individual congruent groups. However, with the inception of whole genome sequencing and comparative genomics, this view has been increasingly blurred. In fact, several characteristics distinguish bacteria and archaea from eukaryotes and these restrict the general application of the biological species concept (Mayr, 1942). Processes that set bacteria and archaea apart include binary fission (Lawrence, 2002), homologous recombination (Fraser *et al.*, 2007) and lateral gene transfer (Fraser *et al.*, 2007). At present, bacteria and archaea are rather arbitrarily grouped into uniform groups based on overall genomic similarity and sharing of phenotypes (Vos, 2011). This polyphasic and pragmatic approach has served the community well but is based on methods rather than being based on cohesive evolutionary forces (Achtman and Wagner, 2008).

Several different models have been proposed to describe microbial (bacterial and archaeal) divergence and speciation. Most of these are based on the *clonal ecotype* model proposed by Cohan and others (Cohan, 2006; Ward, 2006) which in turn is based on the earlier *periodic selection* model (Atwood *et al.*, 1951). According to this model, in an environmental context, a single clonal type occupies a particular niche. This comes about because mutations that lead to increased fitness in the niche periodically arise in the population, leading to selective sweeps and the loss of neutral diversity. Several sequence clusters are inferred to represent occupants of distinct niches (sympatric divergence), or alternatively, the mixing of two physically separated populations (allopatric divergence). Several models have followed from the original model including the *stable ecotype* model, the *geotype* and *Boeing* model, the *genetic drift* model, the *cohesive recombination* model and the *species-less* model (summarized in Gevers *et al.*, 2005). Importantly, the *clonal ecotype* model provides a mechanistic understanding of the evolutionary processes in addition to an organizing principle for classifying species, which is based on solid experimental observations of bacterial populations (Fraser *et al.*, 2009).

Recently, Achtman and Wagner (2008) proposed to apply the method-free unitary species concept of de Queiroz (2005) to microbial systematics. This concept posits that 'species are metapopulation lineages' which extend through time (Achtman and Wagner, 2008). Importantly, in this model, 'metapopulation lineages do not have to be phenotypically distinguishable, or diagnosable, or monophyletic, or reproductively isolated, or ecologically divergent, to be species. They only have to be evolving separately from other such lineages.' Consequently, evolutionary cohesive forces shape these metapopulation lineages and, thus, the concept is not based on methodologies for delineating organismal groups but constitutes an explanation for their existence. Most recently, Vos (2011) has proposed the 'adaptive divergence species concept' in which two phylogenetic clusters are classified as 'separate species only when they are statistically significantly diverged as a result of adaptive evolution', i.e. positive selection. This method-based concept is based on the work of Simmons *et al.* (2008), who applied the McDonald–Kreitman (MK) test to population metagenomic data and found that, based on the MK test, naturally occurring closely related strains of bacteria are most likely not ecologically distinct (see 'Single nucleotide polymorphisms and population genomics' section for a more in depth discussion). Although the concept by Vos (2011) can be broadly applied to genomic and multilocus sequence analysis (MLSA) data, it exhibits a number of limitations and challenges which need to be assessed in detail before it can be universally applied (see Vos, 2011, for a detailed discussion).

Microbial diversification, adaptation and speciation have proven to be 'perennially vexatious questions' (Doolittle and Papke, 2006), and a model that is non-arbitrary, universal and solidly grounded in evolutionary theory (Vos, 2011) has yet to emerge. Such a model should not only provide an integrative view of evolutionary processes leading to microbial divergence but should also reflect ecological niche delineation which in turn will be the functional consequence of genomic differentiation. Furthermore, such a model, through the inclusion of functional details, will be a logical extension of proposed genome-based classification systems, e.g. the phylogenomic species concept (Staley, 2009a,b).

The application of high-resolution molecular tools, including metagenomics, metatranscriptomics, metaproteomics and community metabolomics, offer exciting prospects to obtain integrative views of microbial differentiation over space and time. Natural microbial communities that have proven exceptionally well suited for studying microbial ecology and evolution are biofilms growing within acid mine drainage (AMD) solutions in the Richmond Mine at Iron Mountain (Redding, CA, USA). These biofilms exhibit comparatively low species richness and, thus, can be regarded as ideal model communities to study fundamental ecological and evolutionary processes at the molecular level (Denef *et al.*, 2010b). For example, it was recently demonstrated that integrated proteomics and metabolomics can be used to functionally delineate two closely related microbial species and strains, and determine details of their differing ecological roles (Wilmes *et al.*, 2010; Fischer *et al.*, 2011). Excitingly, the approaches developed on the AMD system now await their application to other mixed microbial communities.

Rather than propose yet another 'microbial species concept', the present chapter reviews current interpretations of high-resolution molecular data and describes how this data reflects processes linked to archaeal and bacterial differentiation, divergence and speciation. Importantly, the integrative application of 'meta-OMIC' methods to a range of microbial communities in the near future, in addition to the placement of the resulting data into ecological and evolutionary context, will lead to molecular definitions of congruent organismal groups, which may provide a sound paradigm for microbial systematics. Since we are only at the start of the meta-OMIC 'revolution' in microbiology, such a paradigm will only emerge once sufficient data has been obtained from various different microbial communities.

Genome-based differentiation

Distinguishing or regrouping different microbial strains according to genomic similarity has been the rationale behind using DNA–DNA hybridization for species classification [strains are assigned to the same species if their reciprocal, pairwise DNA re-association values are ≥ 70% (Wayne *et al.*, 1987)]. The cut-off level for defining microbial species is not based on any particular theoretical justification but was pragmatically chosen to match pre-existing species definitions of microbial isolates in pure culture (Achtman and Wagner, 2008). Although the definition of microbial species based on DNA–DNA re-association values is method-based, it has served microbiologists well so far. However, in particular, the results of microbial genome sequencing efforts have called into question this classification system (Konstantinidis *et al.*, 2006).

In 1985, a new experimental approach based on analysis of small ribosomal subunit RNA (rRNA) genes allowed microbial surveys of different environments to be conducted without the need for cultivation (Handelsman, 2004). This approach revolutionized our understanding of microbial diversity and evolution. More recent studies based on PCR amplification and next-generation high-throughput sequencing of 16S rRNA fragments have revealed vast phylotypic diversity in a wide range of microbial habitats, e.g. ocean (Sogin *et al.*, 2006), soil (Roesch *et al.*, 2007), human beings (Andersson *et al.*, 2008), and air (Tringe *et al.*, 2008). Although these approaches provide important estimates of taxonomic richness within a given community (a cut-off of > 97% 16S rRNA gene sequence similarity is typically used to define operational taxonomic units), they are unable to resolve the true genetic potential contained within regrouped microbial populations and, thus, such studies tell us little about population-level dynamics that may lead to the emergence of new organismal groups. The reason for this disparity is that genome plasticity causes extensive variations in gene content between closely related strains of the same species (Medini *et al.*, 2005). This genomic heterogeneity between strains with slightly divergent (Konstantinidis *et al.*, 2006) or identical (Simmons *et al.*, 2008) 16S rRNA genes complicates evolutionary and functional interpretation of 16S rRNA marker gene profiles. For example, based on DNA re-association kinetics of pooled genomic DNA, Gans and co-workers (2005) estimated that 1 gram of pristine soil may contain 10^6 distinct genotypes. This number far exceeds recent phylotypic diversity estimates for soil, e.g. 52,000 phylotypes (Roesch *et al.*, 2007). Thus, owing to the dynamic nature of microbial genomes, phylotypic diversity clearly does not correlate with genotypic diversity and, hence, genotypic and phenotypic richness within a given sample cannot be inferred from ribosomal RNA surveys alone.

Slight differences in 16S rRNA gene sequences may be linked to resource partitioning and resulting sympatric divergence between closely related strains [<1% 16S rRNA divergence (Hunt *et al.*, 2008)], whereas in other cases such slight differences may be insignificant to functionally differentiate distinct populations, e.g. species (Cilia *et al.*, 1996). Overall, the disparity between phylotypic and genotypic richness does showcase the fact well that using phylotypic profiles to infer population dynamics, differentiation and potentially resulting divergence is inconsistent because phylotypes cannot be placed into congruent 'species bins'. In order to fully appreciate processes related to microbial divergence and speciation, it is thus important to comprehensively resolve the genetic inventories of the respective community members *in situ* and follow these over space and time.

Community genomics (metagenomics) based on high-throughput sequencing of microbial community DNA goes far beyond marker gene surveys (e.g. 16S rRNA gene phylogenies) to provide in-depth insight into the genotypic richness within populations. At the time of writing (July 2011) 307 metagenomic sequencing projects were listed as ongoing in the Genomes OnLine Database (http://www.genomesonline.org). In particular, genome-centric studies (Wilmes *et al.*, 2009b) where extensive *de novo* assembly is obtainable owing to limited species richness (e.g. Tyson *et al.*, 2004), the application of complexity reduction methods (e.g. Pernthaler *et al.*, 2008), or where previously sequenced isolate genomes allow recruitment of genomic fragments (e.g. Coleman *et al.*, 2006) offer exciting prospects to probe the genomic diversity contained within populations, and resolve patterns and mechanisms related to organismal divergence.

Metagenomic data can easily be regrouped into defined sequence clusters following assembly (García-Martín *et al.*, 2006; Hallam *et al.*, 2006; Allen *et al.*, 2007; Eppley *et al.*, 2007b; Rusch *et al.*, 2007; Simmons *et al.*, 2008; Dick *et al.*, 2009; Arumugam *et al.*, 2011) which suggests that some force restricts genetic exchange between populations. Potential explanations include specific adaptation to particular environmental niches among co-existing populations, physical isolation or a decline in recombination frequency between co-existing populations due to neutral divergence within genomes (Wilmes *et al.*, 2009b). Metagenomics has in particular highlighted extensive population-level genetic heterogeneity within mixed microbial communities that mirrors earlier findings in isolate genomes (Wilmes *et al.*, 2009b). From a fundamental point of view, the mechanisms that drive the differentiation of microorganisms (bacteria and archaea) into congruent clusters must restrict the accumulation of neutral diversity (Fraser *et al.*, 2009). However, from observed patterns in high-throughput sequence data, e.g. Simmons *et al.* (2008), it remains unclear if population-level variation is the result of neutral versus adaptive processes (Wilmes *et al.*, 2009b). Consequently, it is helpful to review the molecular patterns of microbial differentiation and place these into mechanistic and theoretical frameworks that may provide a roadmap towards a universal and molecular species definition in the future.

From extensive observations of different bacterial isolates in the laboratory, different mechanisms driving evolution, environmental adaptation and speciation have been identified in bacteria and archaea. Genome-based differentiation of microorganisms may result from (i) random mutations rising to fixation (Wilmes *et al.*, 2009b), (ii) recombination events which may be homologous, i.e. the acquisition of stretches of DNA from closely related microorganisms at flanking homologous sequence stretches (Achtman and Wagner, 2008), or non-homologous, i.e. the acquisition of stretches of DNA from unrelated sources (Achtman and Wagner, 2008), and (iii) epigenetic mechanisms which lead to heritable changes in gene expression through additional information being superimposed on the DNA sequence (Casadesus and Low, 2006; Fig. 1.1).

Single nucleotide polymorphisms and population genomics

The classical model of microbial evolution is the *periodic selection* model, which was based on elegant artificial selection experiments with *E. coli* grown for extended periods under stable conditions in chemostats (Atwood *et al.*, 1951). This work showed repeated selective sweeps and the resulting *periodic selection* model has a long history in bacterial population genetics (Levin, 1981). The model posits that beneficial mutations with large effects on fitness arise rarely in asexual populations but once a clone contains a large effect single nucleotide polymorphism (SNP), this rapidly rises to fixation via a selective sweep. A SNP is a DNA sequence variation that occurs when a single nucleotide differs in the genomes of members of a population. Because recombination (Fraser *et al.*, 2007) and clonal interference (de Visser and Rozen, 2006) are essentially absent in the model, the sweep carries an entire genotype to fixation and eradicates diversity at all other loci. During the period of stasis in between the appearance of large effect mutations, neutral diversity can again accumulate SNPs at multiple loci.

The *periodic selection* model is the basis for the *clonal ecotype* model by Cohan and others

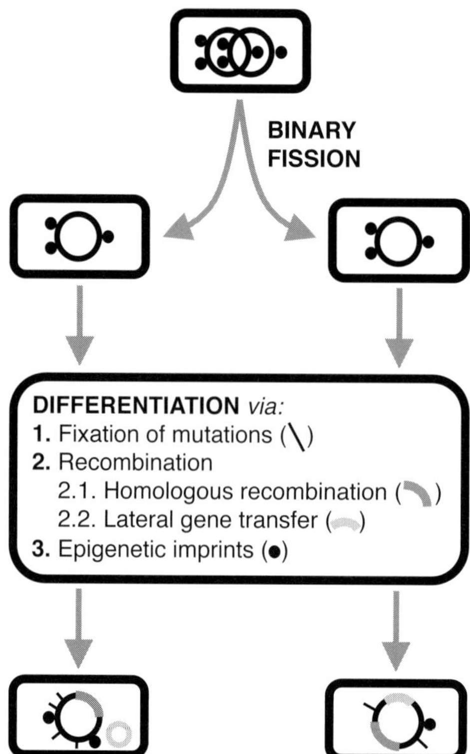

Figure 1.1 Schematic representation of archaeal and bacterial genome-based differentiation mechanisms.

(Cohan, 2006; Ward, 2006). According to this model, in an environmental context, a single clonal type occupies a particular niche. This comes about because mutations that lead to increased fitness in the niche periodically arise in the population, leading to selective sweeps and the loss of neutral diversity. Multiple sequence clusters are inferred to represent occupants of distinct niches, or alternatively, the mixing of two physically separated populations. Typically, one or more marker gene phylogenies are constructed and the clustering of particular phylogenetic groups according to a limited set of environmental parameters is tested. A positive association is interpreted as support for the ecotype model (Ward, 2006; Ward *et al.*, 2007; Koeppel *et al.*, 2008), because distinct sequence clusters correspond to ecologically distinct populations.

Metagenomic data are rich in SNPs but their frequency varies considerably between different organismal groups. SNP frequencies in populations in the AMD system vary from around 0.08% (*Leptospirillum* group II) to 2.2% (*Ferroplasmaacidarmanus;* Tyson *et al.*, 2004), whereas the SNP frequency in four endosymbionts of the marine oligochaete *Olaviusalgarvensis* range from 0.01% (δ4) to 0.1% (γ1; Woyke *et al.*, 2006). However, the calculation of rates of mutation, selection, and recombination within microbial populations is confounded by the inability to physically reconstruct genomes of individual cells, i.e. haplotypes. Genomic contigs produced through automated or manual assembly are composite sequences derived from multiple individuals but statistical reconstructions of individual haplotypes may be possible based on correlations in polymorphism frequency between samples (Wilmes *et al.*, 2009b).

In order to discern evolutionary trajectories from metagenomic data, it is essential to test the adaptive significance of sequence variation within and between populations. This presents a considerable challenge for metagenomic data (Wilmes *et al.*, 2009b). Much of the observed SNPs may be neutral, and persist in microbial populations owing to potentially quite large, but presently unknown, effective population sizes (Mes, 2008). Basic population genetic theory predicts that neutral variation will persist in a population for a number of generations of the same order of magnitude as the effective population size N_e, if genetic drift is the only force acting on it (Gillespie, 2004; Mes, 2008). N_e is defined as the size of a population evolving in the absence of selection that would generate as much neutral diversity as is actually observed (Fraser *et al.*, 2009). N_e is typically smaller than the census population size (N). N_e determines the rate at which variation is lost from a population, and is highly sensitive to bottlenecks (such as periodic selection or predation events). Given the enormous census sizes of microbial populations, however, N_e could still be large enough to ensure an extremely long fixation time for neutral variation. Estimates of N_e for bacteria range from 10^5 to 10^9 (Fraser *et al.*, 2009). However, the number of *Vibrio* cells per m^3 of seawater in temperate coastal regions range from 10^8 to 10^9 which suggests vast census population sizes ($> 10^{20}$; Fraser *et al.*, 2009). The apparent mismatch of several orders of magnitude between N_e and N was used to dispel claims of neutrality implying that all genetic variation is adaptive (Fraser *et al.*, 2009). However, alternative explanations exist (Fraser *et al.*, 2009). For example, one theoretical model suggests that recurrent mutation is sufficient to block the fixation of neutral gene variants in populations with N_e such as seen in bacteria, leading to an enormous, transient flux of novel sequences (Berg and Kurland, 2002). This is consistent with empirical observations of high genotype diversity derived from comparisons of isolates (Thompson *et al.*, 2005) and population genomic assemblies (Allen *et al.*, 2007; Simmons *et al.*, 2008) although other studies seem to contradict this (Hunt *et al.*, 2008; Pena *et al.*, 2010). Although metagenomic data offer exciting prospects for population genetic analyses, the application of existing population genetic tests to these data is not straightforward (Wilmes *et al.*, 2009b). This complicates population genomic analyses in that the estimation of polymorphism frequencies is confounded by the inability to assemble and compare discrete genomes, but this may change in the future with the advent of single-cell genomics.

The determination of selection in individual genes, indels, or intragenic retrieved from population genomic datasets has been applied in a number of studies (Zeidner *et al.*, 2005; Allen *et al.*, 2007; Piganeau and Moreau 2007; Wilhelm *et al.*, 2007). Frequency spectrum and neutral/non-neutral mutations are determined for all genes or a subset of genes that may be interesting from a ecological differentiation point of view (Wilmes *et al.*, 2009b). Initial tests whether the frequency distribution of SNPs in a set of aligned sequences is consistent with positive, negative, or no selection under particular evolutionary models are carried out (Wilmes *et al.*, 2009b). Secondly, a comparison of the number of synonymous substitutions (assumed to be neutral) with the number of non-synonymous substitutions (assumed to have a fitness effect) is also carried out (Wilmes *et al.*, 2009b). In its most basic form, this is used to calculate the ratio of non-synonymous substitutions per non-synonymous site (dN or K_n) to synonymous substitutions per synonymous site (dS or K_s). A dN/dS ratio greater than one is generally assumed to indicate positive selection, because non-synonymous substitutions would not be retained in the population unless they increased individual fitness (Wilmes *et al.*, 2009b). The d*N*/d*S* ratio is particularly relevant because single non-synonymous mutations can alter the kinetics and specificity of enzymes, providing a mechanism for allopatric and sympatric divergence. For example, a single amino acid substitution can switch marine proteorhodopsins (a widely distributed light-driven proton pump) from blue light to green light absorbing (Kelemen *et al.*, 2003; Man *et al.*, 2003) and this point mutation allows spectral tuning according to position along a depth-dependent light gradient and, hence, niche occupation (Beja *et al.*, 2001). In the AMD system, two cytochromes involved in iron oxidation exhibit high levels of sequence variation relative to the rest of the genome of *Leptospirillum* group II (Jeans *et al.*, 2008; Singer *et al.*, 2008, 2010) which may be a reflection of the existence of micro-scale niches along geochemical gradients (Wilmes *et al.*, 2009a; Ma and Banfield, 2011). Most large-scale studies of d*N*/d*S* detect purifying selection on nearly all genes (Allen *et al.*, 2007; Petersen *et al.*, 2007). However, this does not account for the different degrees of adaptive power between genes.

The McDonald–Kreitman test (MK; McDonald and Kreitman, 1991) with refinements (Bustamante *et al.*, 2002) is a more powerful use of counts of synonymous and non-synonymous data than simply calculating the d*N*/d*S* ratio (Wilmes *et al.*, 2009b). As highlighted above, this test forms the basis of the 'adaptive divergence species concept' formulated by Vos (2011). The test posits that under a model of neutral evolution, the ratio of non-synonymous to synonymous substitutions within a population is the same as the ratio of non-synonymous to synonymous fixed differences between populations. An excess of replacement fixed differences indicates positive selection on a particular locus, whereas a dearth indicates negative selection. This test is particularly well suited for testing the 'ecotype' model (Ward *et al.*, 2007). The model is based on the hypothesis that regions differentiating co-existing organisms should encode genes responsible for their increased fitness in specialized niches. If genomic regions are orthologous, and each co-existing organism is uniquely adapted to a particular niche, the MK test should show increased evidence of positive selection in these regions relative to the rest of the genome. So far, the MK test has not been widely applied to metagenomic data (but see Simmons *et al.*, 2008, and discussion below).

The pioneering metagenomic sequencing of biofilms obtained from AMD solutions at the five-way sampling location within the Richmond Mine allowed Tyson *et al.* (2004) to reconstruct near-complete composite genomes for dominant bacteria, i.e. *Leptospirillum* groups II, and archaea, i.e. *Ferroplasma* type II. This initial data, in addition to additional metagenomic sequence data obtained from a separate sampling location, i.e. UBA (Lo *et al.*, 2007), has been used to obtain comprehensive and deeply sampled genomic datasets for multiple organisms and has allowed for a direct analysis of *in situ* population heterogeneity. Interestingly, within a given sample, the level of within-population variability ranges from near-clonal (Lo *et al.*, 2007) to freely recombining (Eppley *et al.*, 2007b). Given these differing patterns, major questions related to microbial divergence can be addressed in these microbial communities simultaneously (Denef *et al.*, 2010b).

The two deepest coverage assemblies were obtained for two *Leptospirillum* group II populations sampled at the UBA and five-way locations within the mine (Tyson *et al.*, 2004; Lo *et al.*, 2007). Based on the size of the pangenome of the *Leptospirillum* group II populations, it was estimated that the number of unique genotypes was only one order of magnitude less than the number of cells in the population (Simmons *et al.*, 2008). The UBA population of *Leptospirillum* group II was dominated by one sequence type with relatively abundant variants (> 99.5% nucleotide identity to the dominant type) also present. Blocks of other *Leptospirillum* group II types (around 94% sequence identity) were shown to have recombined into one or more variant types. The two dominant *Leptospirillum* group II populations from the UBA and five-way locations are approximately 95% identical at the amino acid level, though they have been shown to actively recombine between each other (Lo *et al.*, 2007; Denef *et al.*, 2009, 2010a). Furthermore, there is even more fine-scale recombination within the *Leptospirillum* group II five-way population between distinct substrains that are less than 0.5% divergent (Simmons *et al.*, 2008).

Population genetic analyses of SNPs using the MK test indicated that variation between closely related strains of *Leptospirillum* group II is not maintained by positive selection which in turn means that strain-variant regions do not represent adaptive differences between strains (Simmons *et al.*, 2008). In addition, the majority of *Ferroplasma* type I genes were found to be under strong stabilizing selection; only six loci out of 1963 exhibited non-synonymous versus synonymous SNP ratios indicative of positive selection (Allen *et al.*, 2007). This implies that the ecotype model is not applicable to these populations. However, a major limitation of the application of the MK test is that this metric may obscure biological variation in the degree of adaptive divergence between genes (Vos, 2011). The method is also only applicable to homologous genes, most of which are found in the core genome. At the population-level, because of considerable niche overlap, homologous genes may be evolutionarily constrained, thus biasing the analysis. This may explain discrepancies in the results obtained by Simmons *et al.* (2008), i.e. variation between closely related strains is not maintained by positive selection, whereas more recently, based on functional proteomics data, Denef *et al.* (2010) found evidence for ecological adaptation among *Leptospirillum* group II populations.

The methods developed in the AMD system have been applied to other communities, in particular enriched mixed cultures of '*Candidatus* Accumulibacter phosphatis' (*A. phosphatis*) an organism responsible for the process of enhanced biological phosphorus removal (EBPR) in biological wastewater treatment plants (García-Martín *et al.*, 2006). In this study, the enriched mixed cultures from two different sludges were sequenced, one from a reactor operated in the United States and one operated in Australia. The dominant *A. phosphatis* organisms comprised similar genotypes (> 95% identical at the nucleotide level) but extensive strain diversity was apparent which included divergence of up to 15% at the nucleotide level (García-Martín *et al.*, 2006). The genome-wide patterns observed in the dominant populations within the AMD biofilms as well as those observed in the *A. phosphatis* populations have been investigated further using functional post-genomic approaches (see 'Functional and regulatory differentiation' section below). Overall, the SNP patterns observed in the dominant populations in both systems were rather dissimilar, which is reflected in the suggestion that different genomic differentiation strategies may be employed by microorganisms exposed to different environmental pressures.

Recombination

Bacteria and archaea evolve rapidly not only by processes of random mutation (for which there is extensive evidence in natural populations, 'Single nucleotide polymorphisms and population genomics' section) but also by capture of foreign DNA from different sources. In fact, genetic exchange, once thought to be uncommon in haploid organisms, is now recognized as a major driver of microbial evolution. In particular, metagenomic data have highlighted the importance of both homologous and non-homologous recombination in shaping population structures. Recombination is the process by which a nucleic acid molecule is cleaved and then ligated to a different molecule. Overall, theoretical and experimental work has demonstrated that recombination provides a fitness advantage in microbial populations which is of the same frequency than the mutation rate (Cooper, 2007), suggesting that clonal models alone do not adequately reflect genome-based microbial divergence.

The uptake and use of alien genetic material was first demonstrated in early experiments with *Streptococcus pneumonia* (Griffiths, 1928). Genetic exchange within or between populations is mediated by the three mechanisms of conjugation, transduction and transformation (Thomas and Nielsen, 2005). Conjugation is the transfer of genetic material between microbial cells that requires direct cell-to-cell contact. Transduction is the acquisition of foreign DNA via a viral intermediate. Transformation is the process of scavenging and incorporating DNA from the environment. All three processes leave imprints in microbial genomes and contribute substantially towards the observed population-level genetic heterogeneity. The genes transferred to recipient cells typically form part of the accessory genome whereas the core genome typically remains largely unaffected. The level of genetic exchange and sequence divergence within and between populations varies from little (near-clonal) to high (free recombination), as measured both by population genomic techniques and multilocus sequence analyses (MLSA) of isolates (reviewed by Pérez-Losada *et al.*, 2006).

Missing haplotype information in community genomic data creates particular problems for methods designed to detect recombination through comparisons of sequences from different individuals using coalescent theory (e.g. McVean *et al.*, 2002; Fearnhead *et al.*, 2004) or phylogenetic break-point methods (Minin *et al.*, 2005). Using these methods, recombination can only be detected at length scales smaller than an individual clone (Wilmes *et al.*, 2009b). This circumstance renders any model-based detection of recombination over longer length scales or in less variable genomes cumbersome or impossible (Wilmes *et al.*, 2009b). The only studies to tackle the problem of measuring recombination rates in large-scale population genomic datasets obtained from the AMD system have done so using manual identification of breakpoints, which requires a polymorphism density high enough for visual detection (Whitaker and Banfield, 2006; Eppley *et al.*, 2007b; Simmons *et al.*, 2008;). However, recombination breakpoints can also be identified and confirmed using proteomics inferred genotyping (Lo *et al.*, 2007; Denef *et al.*, 2009), a rapid and high-resolution method.

In the AMD biofilms the level of within-population genomic differences resulting from recombination varies to different degrees. The dominant *Leptospirillum* group II species with its two sequenced dominant strains, i.e. five-way and UBA types, shows signs of homologous recombination (Lo *et al.*, 2007), but this is not nearly as pronounced as for the archaea referred to as *Ferroplasma* type I and II, which have pronounced mosaic genomes (Tyson *et al.*, 2004; Allen *et al.*, 2007). The two *Leptospirillum* group II strains are approximately 95% identical at the amino acid level, however, they do actively recombine (Lo *et al.*, 2007) leading to a blurring of their respective genotypes in some samples (Denef *et al.*, 2009). In addition, there is also evidence for recombination within the *Leptospirillum* group II five-way population between distinct substrains less than 0.5% divergent from the dominant type (Simmons *et al.*, 2008). Putative recombination breakpoints between the very closely related strains of the bacterium were identified with the visualization program

Strainer (Eppley *et al.*, 2007a), but owing to the low overall polymorphism density, the exact location could not be defined conclusively (Simmons *et al.*, 2008). However, based on the extensive gene content variation within the *Leptospirillum* group II populations, it could be inferred that the number of unique genotypes was only one order of magnitude less than the number of cells in the population (Simmons *et al.*, 2008). Remarkably, although the *Leptospirillum* group II populations show extensive recombination, the *Ferroplasma* type II population within the five-way sample consisted of individuals with mosaic genomes formed by recombination between numerous distinct genome types (Tyson *et al.*, 2004). A comparison of a *Ferroplasma* type I isolate with its corresponding populations in the metagenomically characterized sample revealed that much of the observed heterogeneity was due to transposase movement and phage insertions and deletions (Allen *et al.*, 2007). Importantly, recombination was more frequent within both *Ferroplasma* type I and type II populations than between them, consistent with a log-linear decline in recombination frequency with sequence divergence (Eppley *et al.*, 2007b), highlighting the existence of defined and constrained sequence clusters.

In contrast to the extensive patterns of recombination observed within dominant microbial populations in AMD biofilms, sampled *A. phosphatis* strains (García-Martín *et al.*, 2006) do not exhibit evidence of genomic mosaicism or even modest levels of homologous recombination (Kunin *et al.*, 2008). Consequently, recombination frequency may be species or strain specific, and may be linked to the intensity of selective pressures in a given environment.

The adaptive importance of variable genomic regions in an environmental context is convoluted. Laboratory experiments with isolated strains of *Prochlorococcus* spp. support the importance of hypervariable regions in environmental adaptation (Coleman *et al.*, 2006). In one *Prochlorococcus* strain, 26% of all genes in highly variable regions of the genome were differentially expressed under changed nutrient or light conditions in culture (Coleman *et al.*, 2006). Specific variable gene clusters in *Prochlorococcus* reflect phosphorus scarcity in the Atlantic Ocean (Coleman and Chisholm, 2010). Furthermore, genomic islands in *Prochlorococcus* have also recently been implicated in resistance to viral predation (Avrani *et al.*, 2011), a major selective pressure to which microorganisms are exposed (Wilmes *et al.*, 2009b). Bioinformatic analysis also supports the potential adaptive value of genomic islands in other species. For example, several of the genomic islands differentiating the soil bacterium *Burkholderia xenovorans* LB400 from other strains of its species contain the genes that enable it to degrade chlorinated aromatics (Chain *et al.*, 2006). Many additional examples regarding the importance of genomic islands have been summarized elsewhere (e.g. Dobrindt *et al.*, 2004).

Virus–host interactions may have a significant effect on genomic adaptation within communities. Viruses are essential mediators of genetic exchange in the environment (Ripp *et al*, 1994; Jiang and Paul, 1998). As agents of gene transfer, viruses may supply the host with new genetic material in the form of integrated elements (reviewed by Faruque and Mekalanos, 2003; Sherwood, 2003; Brussow *et al.*, 2004) and replace cellular genes by viral non-orthologues (Filée *et al.*, 2002, 2003). In some cases, viral genes are known to increase the survival fitness of the host (Brussow *et al.*, 2004). For example, cyanophages infecting *Synechococcus* and *Prochlorococcus* carry genes involved in photosynthesis (Mann *et al.*, 2003; Lindell *et al.*, 2004). Expression of cyanophage-encoded photosystem proteins (psbA/psbD) helps to support photosynthetic activity in the host during the infection cycle, providing photosynthetic gene carrying cyanophages with a selective advantage (Lindell *et al.*, 2004). Viral *psbA* and *psbD* have been detected in open ocean metagenomic surveys (Venter *et al.*, 2004; Angly *et al.*, 2006; DeLong *et al.*, 2006; Rusch *et al.*, 2007). Viral *psbA* genes are evolving under levels of purifying selection that are virtually indistinguishable from those acting on host proteins (Zeidner *et al.*, 2005). There is also evidence for exchange and reshuffling of *psbA* genes between *Synechococcus* and *Prochlorococcus* via phage intermediates, as well as between phages and hosts and between phages (Sullivan *et al.*, 2003). Consequently, viruses appear to play a role in both short- and long-term adaptation in host populations but, at present, little is known about molecular mechanisms facilitating rapid genome evolution in microbial viruses. Comparative genomics suggest that the viral gene pool appears to be shaped primarily by illegitimate and homologous recombination (Hendrix, 2003; Martinsohn *et al.*, 2008). Apart from recombination, diversity-generating retro-elements (Liu *et al.*, 2002) allow viruses to generate adaptive diversity through a stochastic mechanism.

Recombination unlinks the evolutionary fate of different parts of a genome, allowing selection to operate independently on individual loci or sets of linked loci (Wilmes *et al.*, 2009b). If natural selection is relatively weak, the net effect is higher levels of standing diversity within a population than we would expect from clonal models alone (Wilmes *et al.*, 2009b). If recombination in a population is extensive, phylogenetic signals of descent can be obscured (Wilmes *et al.*, 2009b). In fact, incongruence between phylogenies derived from different loci within a population is a widely used indicator for the occurrence of recombination (Feil and Spratt, 2001). The combination of recombination and weak selection can therefore result in the appearance of sequence clusters that do not correspond to ecologically unique species (Cohan, 2006; Whitaker and Banfield, 2006). An important emerging pattern is that the rates of intra-population recombination exceed that of inter-population recombination. The latter observation may be explained by the incompatibility of distantly related DNA with the recipient's restriction modification systems, its immune system, in particular the clustered regularly interspaced short palindromic repeats (CRISPR) system (Wilmes *et al.*, 2009b), as well as its replication, transcription and translation machinery. All of these mechanisms will lead to incompatible stretches of DNA being rapidly purged from microbial populations.

Epigenetic mechanisms

As highlighted above, community genomics provides an overview of the prevalence of random mutations and recombination events within natural microbial populations which may be a reflection of ecological adaptation and microbial divergence/speciation among distinct genotypes. However, mutation and recombination may not be the only mechanisms at play in natural microbial populations that result in phenotypic differentiation. Early laboratory-based cultivation studies revealed that bacterial populations harbour an inconsistent number of bacteriophage-resistant individuals (Luria and Delbrück, 1943). Resistance was related to the activity of restriction-modification systems (Wilson and Murray, 1991), which comprise a methyltransferase that modifies adenine or cytosine residues at certain recognition sites and a restriction endonuclease that cleaves the DNA at the recognition site if it is unmethylated. These systems are especially useful for protecting microbial cells against the unregulated uptake of foreign DNA and, thus, viral defence. The superimposition of additional information onto genomic DNA sequences through methylation is considered 'epigenetic' (Casadesus and Low, 2006). Epigenetic mechanisms are diverse but most epigenetic systems known in bacteria use DNA methylation as a signal that regulates a specific DNA–protein interaction (Casadesus and Low, 2006). DNA adenine methylation-dependent regulatory systems control bacterial gene expression and the associated epigenetic regulatory methylases are designated as orphan because they lack a cognate restriction enzyme (Casadesus and Low, 2006). Restriction-modification systems are very common in bacteria and differential epigenetic regulation of gene expression may facilitate rapid adaptation to different ecological niches. Consequently, differential methylation imprints may be an essential starting point for speciation (Jeltsch, 2003).

Although the existence and activity of methylation-based epigenetic mechanisms is very well established in bacteria, large-scale (meta-)epigenomic sequencing studies have so far not been conducted because bacteria use DNA adenine methylation rather than cytosine methylation as an epigenetic signal. However, the advent of single-nucleotide, real-time sequencing in combination with polymerase kinetics monitoring is poised to change this (Flusberg *et al.*, 2010). Population-level methylation patterns will provide clues about the importance of epigenetic differentiation in microbial speciation.

Functional and regulatory differentiation

Genetic heterogeneity is extensive within mixed microbial communities but population genomic analyses are inconclusive with respect to the significance of this variation. Post-genomic approaches have the potential to inform whether certain variants or genotypes are functionally relevant and whether the observed genomic variation is the result of neutral versus adaptive processes. Furthermore, differential gene expression detected either at the RNA or protein level may indicate differential gene regulation, which may be essential for rapid (sub-generation timescale) environmental adaptation.

Functional insights into microbial divergence

The use of functional 'omic' data for delineating closely related organisms was demonstrated by Dopson and co-workers (Dopson *et al.*, 2004) who showed that differing protein expression patterns on two-dimensional polyacrylamide gel electrophoresis profiles could be employed to classify four phenotypically distinct *Ferroplasma* spp. isolates obtained from AMD biofilms that were > 98.9% similar at the 16S rRNA gene level with one isolate being borderline on the 70% species boundary from DNA–DNA re-association data. Based on profile similarities an un-rooted tree was constructed, which was congruent with a tree derived from DNA–DNA similarities. Consequently, based on protein expression patterns alone, closely related organisms can be distinguished at least at the species level. However, sub-species population-level differentiation is somewhat more challenging.

In AMD biofilm communities, population genomic studies so far do not support large fitness effects for regions of variable gene content (Simmons *et al.*, 2008). The two main *Leptospirillum* group II genotypes, i.e. five-way and UBA types, dominating the Richmond Mine AMD system differ by only 0.3% at the 16S rRNA gene level, and 20% of each organism's genome is unique relative to the other (Lo *et al.*, 2007). An extensive analysis of 27 environmental proteomes derived from biofilm samples taken from a variety of environmental conditions revealed that recombination between both types is an important mechanism for fine-scale adaptation in the AMD biofilm system (Denef *et al.*, 2009). Interestingly, protein abundance patterns of *Leptospirillum* group II were most highly and significantly correlated with community composition (organismal membership) rather than geochemical parameters such as pH and temperature (Mueller *et al.*, 2010). Furthermore, *Leptospirillum* group II's physiological state changes according to community maturity, which is necessary for adapting to increased competition within the communities (Mueller *et al.*, 2011). This highlights the ecological importance of community composition in niche delineation. Comparative proteogenomic analyses across the 27 samples demonstrated that differential expression of shared genes and expression of a small subset of the 15% genes unique to each genotype are in particular involved in niche partitioning highlighting the fact that subtle genetic variations can lead to distinct ecological strategies (Denef *et al.*, 2010a).

In line with the recent findings within the AMD system, an earlier community proteomic investigation of *A. phosphatis*-dominated microbial communities cultivated in the UK also found evidence for differential expression of protein variants encoded by closely related strains (Wilmes *et al.*, 2008a,b; Wilmes and Bond, 2009). Detection limits for community proteomics suggest that each variant is present at an abundance of at least a few per cent of the total population (VerBerkmoes *et al.*, 2009). Because recombination is essentially absent in these organisms (Kunin *et al.*, 2008), variant proteins must be the result of random mutations that have risen to fixation. Overall, 59% of the most abundant protein variants were derived from flanking *A. phosphatis* populations and not from the dominant *A. phosphatis* strain in the sequenced sludges. A significant subset of these was involved in core metabolism and EBPR-specific pathways, suggesting an essential role for genetic diversity in maintaining the stable performance of the process. Furthermore, this suggests that distinct *A. phosphatis* subpopulations or strains occupy distinct niches in the heterogeneous biological wastewater treatment environment (Wilmes *et al.*, 2008a).

In addition to differences in protein expression among closely related organisms, organisms may also employ their accessory genome more comprehensively. For example, analysis of environmental transcripts extracted from a marine sample showed that a majority of genes forming part of the flexible gene content of *Prochlorococcus* genomes were both present and expressed at similar levels to core genes (Frias-Lopez *et al.*, 2008). However, although certain genes encoded on mobile genetic elements confer selective advantages, this clearly is not always the case and the functional role of these will be assessed on a case by case basis using OMIC techniques.

Apart from differentiating closely related populations that may be considered to be part of the same species, e.g. *Leptospirillum* group II and *A. phosphatis* populations, functional omic data may also inform whether accepted species boundaries are warranted and whether they reflect ecological divergence. In a recent study, tandem high-throughput proteomics and metabolomics were employed to functionally characterize AMD biofilms (Wilmes *et al.*, 2010). Dominant species referred to as *Leptospirillum* group II and III typically co-occur across all biofilm successional stages but are hypothesized to occupy distinct ecological niches based on their respective genomic complements (Goltsman *et al.*, 2009) and distribution patterns *in situ* (Wilmes *et al.*, 2009a). It was assumed that the two contrasting organismal groups must be discernible from the high-resolution molecular data. In particular, the metabolomics data were hypothesized to reflect the organisms discrete functional roles in the community because the metabolome represents the output that results from the cellular interactions of the genome, transcriptome and proteome and, thus, should be the most sensitive indicator of differing cellular activities between the two species. Deconvolution of the high-resolution molecular data indeed demonstrated that identified proteins and detected metabolites exhibit strong correlation patterns that represent species-specific clusters. Because metabolites are a direct reflection of metabolism which in turn may be regarded as a proxy for resource usage, the metabolomics data was used to assess the niche characteristics of both organisms (Fig. 1.2). Niche breadth for the two *Leptospirillum* species was estimated by determining the degree of similarity between the frequency distribution of metabolite features associated with each of the represented microbial community members using the proportional similarity index according to Feinsinger *et al.* (1981):

$$PS = 1 - 0.5 \sum_i |p_i - q_i|$$

where p_i is the proportion of resource items (metabolites) in state i out of all items used by the population (organismal group), and q_i is the proportion of i items in the resource base available to the population (all metabolite features considered in the analysis).

From this analysis, both *Leptospirillum* species exhibit medial niche breadths

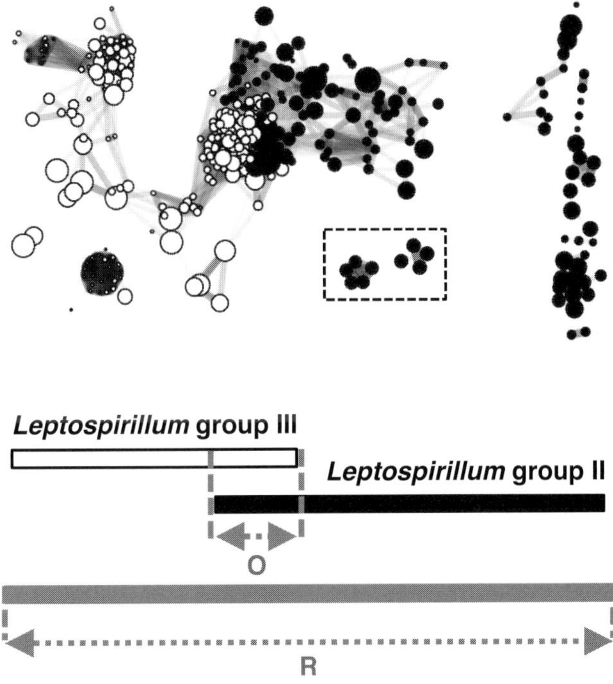

Figure 1.2 Molecular niche characterization through metabolomics. Niche breadths are indicated by bars for both organismal groups, i.e. *Leptospirillum* groups II and III. R designates the entire resource space available to the community, O designates niche overlap between the dominant organisms which in turn governs competition. Sub-network highlighted by dashed rectangle contains *Leptospirillum* group II UBA-type specific lipids. Metabolite correlation network recoloured from (Wilmes *et al.*, 2010).

(proportional similarity indexes of 0.48 and 0.51 for *Leptospirillum* groups II and III, respectively), which indicates a similar degree of specialization for both species. Limited niche overlap (O) suggests partitioned resource usage for both organism groups and the degree of the niche overlap governs competition. Using the same data, competition coefficients ($\alpha_{x(y)}$) were calculated according to (Levins, 1968):

$$\alpha_{x(y)} = \frac{\sum_i \left(p_{xi} p_{yi}\right)}{\sum_i p_{xi}^2}$$

where p_i is again the proportion of resource items (metabolites) in state *i* out of all items (all metabolites considered in the analysis) used by the population (organismal group). In this case, *x* corresponds to *Leptospirillum* group II and *y* to *Leptospirillum* group III, respectively.

The competition coefficients for both *Leptospirillum* groups II and III based on the metabolomics data were found to be very low (α-values of 0.009 and 0.01, respectively). This indicates very little interspecific competition between both organism groups. Because *Leptospirillum* groups II and III are present throughout individual biofilm developmental stages as well as different geochemical conditions, they may be regarded as generalist species. Lack of interspecific competition among generalists is linked to increased individual-level variation and specialization (Bolnick *et al.*, 2007). Interestingly, the observed extensive within-population genetic heterogeneity observed within *Leptospirillum* spp. populations may be a reflection of such adaptive radiation, but this would imply that the population-level genetic heterogeneity is not neutral.

In addition to the species-specific clustering of metabolites, highly abundant metabolites were found to be located in a congruent subnetwork (Fig. 1.2; Fischer *et al.*, 2011). These metabolites have been identified as *N*-monomethyl*lyso*phosphatidylethanolamine lipids. The unusual polar head group structure of these lipids is similar to lipids found in phylogenetically unrelated acidophilic chemoautolithotrophs, e.g. *Acidothiobacillus* spp. and *Halothiobacillus* spp., and this may be related to the affinity of these lipids for iron and calcium ions which in turn would suggest a essential role for these lipids in iron dissolution within the AMD system (Fischer *et al.*, 2011). Based on correlations with strain-resolved protein abundances, these metabolites most likely originate in the UBA-type strain of *Leptospirillum* group II (Fischer *et al.*, 2011). In addition, a putative methyltransferase involved in the conversion of *lyso*phosphatidylethanolamine to *N*-monomethyl *lyso* phosphatidylethanolamine was identified by proteomics and was encoded solely by *Leptospirillum* group II UBA-type (Fischer *et al.*, 2011). This study elegantly demonstrates how closely related strains may become ecologically distinct by encoding strain-specific enzymes that determine their ability to synthesize ecologically relevant small molecules. Overall, the *in situ* species- and strain-level delineation based on metabolomics data are congruent with findings in closely related isolate cultures (Rossello-Mora *et al.*, 2008; Pena *et al.*, 2010).

Regulatory mechanisms and differentiation

In laboratory isolates, divergence has been linked to changes in regulatory circuits resulting, for example, from mutations in highly pleiotrophic regulatory genes (Wang *et al.*, 2010). The observed patterns of species- and strain-specific expression of orthologous proteins suggests that regulatory mechanisms must play an essential role in microbial divergence and differentiation. Denef *et al.* (2010) found, among the proteins suggestive of ecological divergence between the two *Leptospirillum* group II types, numerous putative *S*-adenosyl methionine-dependent methyltransferases. These were found to have an increased role in the late colonizer type (five-way type), which may point towards methylation-dependent epigenetic imprinting being essential for the observed difference in protein expression (discussed above in 'Epigenetic mechanisms' section). Overall, the observed patterns point towards clear differences in protein expression which may be transcriptionally or post-transcriptionally regulated. A lack of clear differences in protein abundance between anaerobic and aerobic cycling of *A. phosphatis*-dominated communities (Wilmes and Bond, 2004, 2006; Wilmes *et al.*, 2008a; Wexler *et al.*, 2009) point towards the fact that post-translational modifications and allosteric regulation may confer a selective advantage in this system (Wilmes *et al.*, 2008a). Finally, regulation by small RNAs is emerging as a key regulatory mechanism in mixed microbial communities (Shi *et al.*, 2009). Overall, differential regulation of transcription, translation and enzyme activities may contribute towards identical genotypes having distinct phenotypes and potentially being recognized as distinct biological entities. The relative importance of the different regulatory mechanisms in organismal differentiation will become apparent with future coupled meta-OMIC studies of natural microbial consortia.

Outlook

We are living through exciting times in microbiology. The application of high-throughput molecular biology methods to natural microbial communities is profoundly changing our view on the microbial world. Since the publication of the first large-scale metagenomic sequencing projects in 2004 (Tyson *et al.*, 2004; Venter *et al.*, 2004), our view of processes leading to microbial divergence and speciation have been profoundly altered. Although potentially adaptive processes, e.g. generation of genetic heterogeneity through random mutations and recombination, had been identified in laboratory isolates, we were until very recently oblivious to the importance of such mechanisms in the natural world.

Although molecular mechanisms that underlie microbial differentiation, divergence and speciation have been identified, interpretation of the relative significance of these processes in a natural context remains challenging. A, if not the, major unresolved question is how much of the observed within-population genetic variation is due to neutral versus adaptive processes. Limited experimental data hints that some of this fine-scale variation being in part functionally relevant (Frias-Lopez *et al.*, 2008; Wilmes *et al.*, 2008a; Denef *et al.*, 2010a), whereas sequence-based and modelling analyses suggest that much of it is neutral (Simmons *et al.*, 2008). Answers to this question are particularly pertinent to explain the emergence of distinct genomic sequence clusters that reflect previous taxonomic definitions of bacterial and archaeal species, e.g. Dick *et al.* (2009).

From the limited amount of largely disjointed metagenomic and functional data obtained to date, extensive intra- and inter-system as well as extensive intra- and inter-population differences are apparent in relation to genetic heterogeneity. Consequently, it is at present difficult to ascertain specific molecular mechanisms and patterns that reflect and define specific microbial groups that would be congruent with a microbial species concept. Concomitant analysis of community genomic complements, transcriptomes, proteomes and metabolomes over relevant spatial and temporal scales will result in detailed molecular descriptions of distinct microbial entities. Such a systems-level molecular organismal classification system will need to be solidly grounded in ecological theory (in particular niche theory), population genetic theory and evolutionary theory (balancing Darwinian and Lamarckian evolutionary processes), and may be more broadly applicable to encompass the three domains of life.

Acknowledgements

This work was supported by a Luxembourg National Research Fund ATTRACT grant awarded to PW (ATTRACT/A09/03). Apologies to authors within the field not included in this review, owing to space limitations literature cited is not exhaustive.

References

Achtman, M., and Wagner, M. (2008). Microbial diversity and the genetic nature of microbial species. Nat. Rev. Microbiol. *6*, 431–440.

Allen, E.E., Tyson, G.W., Whitaker, R.J., Detter, J.C., Richardson, P.M., and Banfield, J.F. (2007). Genome dynamics in a natural microbial strain population. Proc. Natl. Acad. Sci. U.S.A. *104*, 1883–1888.

Andersson, A.F., Lindberg, M., Jakobsson, H., Bäckhed, F., Nyrén, P.l., and Engstrand, L. (2008). Comparative analysis of human gut microbiota by barcoded pyrosequencing. PLoS ONE *3*, e2836.

Angly, F.E., Felts, B., Breitbart, M., Salamon, P., Edwards, R.A., Carlson, C., Chan, A.M., Haynes, M., Kelley, S., Liu, H., *et al.* (2006). The marine viromes of four oceanic regions. PLoS Biol. *4*, e368.

Arumugam, M., Raes, J., Pelletier, E., Le Paslier, D., Yamada, T., Mende, D.R., Fernandes, G.R., Tap, J., Bruls, T., Batto, J.-M., *et al.* (2011). Enterotypes of the human gut microbiome. Nature *473*, 174–180.

Atwood, K.C., Schneider, L.K., and Ryan, F.J. (1951). Periodic selection in *Escherichia coli*. Proc. Natl. Acad. Sci. U.S.A. *37*, 146–155.

Avrani, S., Wurtzel, O., Sharon, I., Sorek, R., and Lindell, D. (2011). Genomic island variability facilitates *Prochlorococcus*-virus coexistence. Nature *474*, 604–608.

Beja, O., Spudich, E.N., Spudich, J.L., Leclerc, M., and DeLong, E.F. (2001). Proteorhodopsin phototrophy in the ocean. Nature *411*, 786–789.

Berg, O.G., and Kurland, C.G. (2002). Evolution of microbial genomes: sequence acquisition and loss. Mol. Biol. Evol. *19*, 2265–2276.

Bolnick, D.I., Svanbäck, R., Araújo, M.S., and Persson, L. (2007). Comparative support for the niche variation hypothesis that more generalized populations also are more heterogeneous. Proc. Natl. Acad. Sci. U.S.A. *104*, 10075–10079.

Brussow, H., Canchaya, C., and Hardt, W.-D. (2004). Phages and the evolution of bacterial pathogens: from genomic rearrangements to lysogenic conversion. Microbiol. Molec. Biol. Rev. *68*, 560–602.

Bustamante, C.D., Nielsen, R., Sawyer, S.A., Olsen, K.M., Purugganan, M.D., and Hartl, D.L. (2002). The cost of inbreeding in Arabidopsis. Nature *416*, 531–534.

Casadesus, J., and Low, D. (2006). Epigenetic gene regulation in the bacterial world. Microbiol. Mol. Biol. Rev. *70*, 830–856.

Chain, P.S.G., Denef, V.J., Konstantinidis, K.T., Vergez, L.M., Agullo, L., Reyes, V.L., Hauser, L., Cordova, M., Gomez, L., Gonzalez, M., *et al.* (2006). *Burkholderia xenovorans* LB400 harbors a multi-replicon, 9.73-Mbp genome shaped for versatility. Proc. Natl. Acad. Sci. U.S.A. *103*, 15280–15287.

Cilia, V., Lafay, B., and Christen, R. (1996). Sequence heterogeneities among 16S ribosomal RNA sequences, and their effect on phylogenetic analyses at the species level. Mol. Biol. Evol. *13*, 451–461.

Cohan, F. (2006). Towards a conceptual and operational union of bacterial systematics, ecology, and evolution. Phil. Trans. Roy. Soc. B: Biol. Sci. *361*, 1985–1996.

Cohan, F.M. (2002). What are bacterial species? Ann. Rev. Microbiol. *56*, 457–487.

Coleman, M.L., and Chisholm, S.W. (2010). Ecosystem-specific selection pressures revealed through comparative population genomics. Proc. Natl. Acad. Sci. U.S.A. *107*, 18634–18639.

Coleman, M.L., Sullivan, M.B., Martiny, A.C., Steglich, C., Barry, K., DeLong, E.F., and Chisholm, S.W. (2006). Genomic islands and the ecology and evolution of *Prochlorococcus*. Science *311*, 1768–1770.

Cooper, T.F. (2007). recombination speeds adaptation by reducing competition between beneficial mutations in populations of *Escherichia coli*. PLoS Biol. *5*, e225.

DeLong, E.F., Preston, C.M., Mincer, T., Rich, V., Hallam, S.J., Frigaard, N.-U., Martinez, A., Sullivan, M.B., Edwards, R., Brito, B.R., *et al.* (2006). Community genomics among stratified microbial assemblages in the Ocean's interior. Science *311*, 496 – 503.

Denef, V.J., VerBerkmoes, N.C., Shah, M.B., Abraham, P., Lefsrud, M., Hettich, R.L., and Banfield, J.F. (2009). Proteomics-inferred genome typing (PIGT) demonstrates inter-population recombination as a strategy for environmental adaptation. Environ. Microbiol. *11*, 313–325.

Denef, V.J., Kalnejais, L.H., Mueller, R.S., Wilmes, P., Baker, B.J., Thomas, B.C., VerBerkmoes, N.C., Hettich, R.L., and Banfield, J.F. (2010a). Proteogenomic basis for ecological divergence of closely related bacteria in natural acidophilic microbial communities. Proc. Natl. Acad. Sci. U.S.A. *107*, 2383–2390.

Denef, V.J., Mueller, R.S., and Banfield, J.F. (2010b). AMD biofilms: using model communities to study microbial evolution and ecological complexity in nature. ISME J. *4*, 599–610.

Dick, G.J., Andersson, A.F., Baker, B.J., Simmons, S.L., Thomas, B.C., Yelton, A.P., and Banfield, J.F. (2009). Community-wide analysis of microbial genome sequence signatures. Genome Biol. *10*, R85.

Dobrindt, U., Hochhut, B., Hentschel, U., and Hacker, J. (2004). Genomic islands in pathogenic and environmental micro-organisms. Nat. Rev. Micro. *2*, 414–424.

Doolittle, W.F., and Papke, R.T. (2006). Genomics and the bacterial species problem. Genome Biol. *7*, 116.

Dopson, M., Baker-Austin, C., and Bond, P.L. (2004). First use of two-dimensional polyacrylamide gel electrophoresis to determine phylogenetic relationships. J. Microbiol. Methods *58*, 297–302.

Eppley, J., Tyson, G., Getz, W., and Banfield, J. (2007a). Strainer: software for analysis of population variation in community genomic datasets. BMC Bioinformatics *8*, 398.

Eppley, J.M., Tyson, G.W., Getz, W.M., and Banfield, J.F. (2007b). Genetic exchange across a species boundary in the archaeal genus Ferroplasma. Genetics *177*, 407–416.

Faruque, S.M., and Mekalanos, J.J. (2003). Pathogenicity islands and phages in *Vibrio cholerae* evolution. Trends Microbiol. *11*, 505–510.

Fearnhead, P., Harding, R.M., Schneider, J.A., Myers, S., and Donnelly, P. (2004). Application of coalescent methods to reveal fine-scale rate variation and recombination hotspots. Genetics *167*, 2067–2081.

Feil, E.J., and Spratt, B.G. (2001). Recombination and the population structures of bacterial pathogens. Annu. Rev. Microbiol. *55*, 561–590.

Feinsinger, P., Spears, E.E., and Poole, R.W. (1981). A simple measure of niche breadth. Ecology *62*, 27–32.

Filée, J., Forterre, P., Sen-Lin, T., and Laurent, J. (2002). Evolution of DNA polymerase families: evidences for multiple gene exchange between cellular and viral proteins. J. Mol. Evol. *54*, 763–773.

Filée, J., Forterre, P., and Laurent, J. (2003). The role played by viruses in the evolution of their hosts: a view based on informational protein phylogenies. Res. Microbiol. *154*, 237–243.

Fischer, C.R., Wilmes, P., Bowen, B.P., Northen, T.R., and Banfield, J.F. (2011). Deuterium-exchange metabolomics reveals N-methyl *lyso* phosphatidylethanolamines as abundant lipids in acidophilic mixed microbial communities. Metabolomics *1*, 13.

Flusberg, B.A., Webster, D.R., Lee, J.H., Travers, K.J., Olivares, E.C., Clark, T.A., Korlach, J., and Turner, S.W. (2010). Direct detection of DNA methylation during single-molecule, real-time sequencing. Nat. Methods *7*, 461–465.

Fraser, C., Hanage, W.P., and Spratt, B.G. (2007). Recombination and the nature of bacterial speciation. Science *315*, 476–480.

Fraser, C., Alm, E.J., Polz, M.F., Spratt, B.G., and Hanage, W.P. (2009). the bacterial species challenge: Making sense of genetic and ecological diversity. Science *323*, 741–746.

Frias-Lopez, J., Shi, Y., Tyson, G.W., Coleman, M.L., Schuster, S.C., Chisholm, S.W., and DeLong, E.F. (2008). Microbial community gene expression in ocean surface waters. Proc. Natl. Acad. Sci. U.S.A. *105*, 3805–3810.

Gans, J., Wolinsky, M., and Dunbar, J. (2005). Computational improvements reveal great bacterial diversity and high metal toxicity in soil. Science *309*, 1387–1390.

García-Martín, H., Ivanova, N., Kunin, V., Warnecke, F., Barry, K.W., McHardy, A.C., Yeates, C., He, S., Salamov, A.A., Szeto, E., *et al.* (2006). Metagenomic analysis of two enhanced biological phosphorus removal (EBPR) sludge communities. Nature Biotech. *24*, 1263–1269.

Gevers, D., Cohan, F.M., Lawrence, J.G., Spratt, B.G., Coenye, T., Feil, E.J., Stackebrandt, E., de Peer, Y.V., Vandamme, P., Thompson, F.L., *et al.* (2005). Re-evaluating prokaryotic species. Nat. Rev. Micro. *3*, 733–739.

Gillespie, J. (2004). Population Genetics: A Concise Guide, 2nd edn (The Johns Hopkins University Press, Baltimore, MD).

Goltsman, D.S.A., Denef, V.J., Singer, S.W., VerBerkmoes, N.C., Lefsrud, M., Mueller, R.S., Dick, G.J., Sun, C.L., Wheeler, K.E., Zemla, A., *et al.* (2009). Community genomic and proteomic analyses of chemoautotrophic iron-oxidizing "Leptospirillum rubarum" (Group II) and "*Leptospirillum ferrodiazotrophum*" (Group III) bacteria in acid mine drainage biofilms. Appl. Environ. Microbiol. *75*, 4599–4615.

Griffiths, F. (1928). The significance of pneumococcal types. J. Hygiene *27*, 113–159.

Hallam, S.J., Konstantinidis, K.T., Putnam, N., Schleper, C., Watanabe, Y., Sugahara, J., Preston, C., de la Torre, J., Richardson, P.M., and DeLong, E.F. (2006). Genomic analysis of the uncultivated marine crenarchaeote Cenarchaeum symbiosum. Proc. Natl. Acad. Sci. U.S.A. *103*, 18296–18301.

Handelsman, J. (2004). Metagenomics: application of genomics to uncultured micro-organisms. Microbiol. Molec. Biol. Rev. *68*, 669–685.

Hendrix, R.W. (2003). Bacteriophage genomics. Curr. Opin. Microbiol. *6*, 506–511.

Hunt, D.E., David, L.A., Gevers, D., Preheim, S.P., Alm, E.J., and Polz, M.F. (2008). Resource partitioning and sympatric differentiation among closely related bacterioplankton. Science *320*, 1081–1085.

Jeans, C., Singer, S.W., Chan, C.S., VerBerkmoes, N.C., Shah, M., Hettich, R.L., Banfield, J.F., and Thelen, M.P. (2008). Cytochrome 572 is a conspicuous membrane protein with iron oxidation activity purified directly from a natural acidophilic microbial community. ISME J. *2*, 542–550.

Jeltsch, A. (2003). Maintenance of species identity and controlling speciation of bacteria: a new function for restriction/modification systems? Gene *317*, 13–16.

Jiang, S.C., and Paul, J.H. (1998). Gene transfer by transduction in the marine environment. Appl. Environ. Microbiol. *64*, 2780–2787.

Kelemen, B.R., Du, M., and Jensen, R.B. (2003). Proteorhodopsin in living color: diversity of spectral properties within living bacterial cells. Biochim. Biophys. Acta (BBA) – Biomembranes *1618*, 25–32.

Koeppel, A., Perry, E.B., Sikorski, J., Krizanc, D., Warner, A., Ward, D.M., Rooney, A.P., Brambilla, E., Connor, N., Ratcliff, R.M., *et al.* (2008). Identifying the fundamental units of bacterial diversity: A paradigm shift to incorporate ecology into bacterial systematics. Proc. Nat. Acad.Sci. USA *105*, 2504–2509.

Konstantinidis, K.T., Ramette, A., and Tiedje, J.M. (2006). The bacterial species definition in the genomic era. Philos. Trans. R Soc. Lond. B Biol. Sci. *361*, 1929–1940.

Kunin, V., He, S., Warnecke, F., Peterson, S.B., Garcia Martin, H., Haynes, M., Ivanova, N., Blackall, L.L., Breitbart, M., Rohwer, F., *et al.* (2008). A bacterial metapopulation adapts locally to phage predation despite global dispersal. Genome Res. *18*, 293–297.

Lawrence, J.G. (2002). Gene transfer in bacteria: Speciation without species? Theoret. Pop. Biol. *61*, 449–460.

Levin, B.R. (1981). Periodic selection, infectious gene exchange and the genetic structure of *E. coli* populations. Genetics *99*, 1–23.

Levins, R. (1968). Evolution in Changing Environments: Some Theoretical Explorations (Princeton University Press, Princeton, NJ).

Lindell, D., Sullivan, M.B., Johnson, Z.I., Tolonen, A.C., Rohwer, F., and Chisholm, S.W. (2004). Transfer of photosynthesis genes to and from *Prochlorococcus* viruses. Proc. Natl. Acad. Sci. U.S.A. *101*, 11013–11018.

Liu, M., Deora, R., Doulatov, S.R., Gingery, M., Eiserling, F.A., Preston, A., Maskell, D.J., Simons, R.W., Cotter, P.A., Parkhill, J., *et al.* (2002). Reverse transcriptase-mediated tropism switching in Bordetella bacteriophage. Science *295*, 2091–2094.

Lo, I., Denef, V.J., VerBerkmoes, N.C., Shah, M.B., Goltsman, D., DiBartolo, G., Tyson, G.W., Allen, E.E., Ram, R.J., Detter, J.C., *et al.* (2007). Strain-resolved community proteomics reveals recombining genomes of acidophilic bacteria. Nature *446*, 537–541.

Luria, S.E., and Delbrück, M. (1943). Mutation of bacteria from virus sensitivity to virus resistance. Genetics *28*, 491–511.

Ma, S., and Banfield, J.F. (2011). Micron-scale Fe^{2+}/Fe^{3+}, intermediate sulfur species and O_2 gradients across the biofilm–solution–sediment interface control biofilm organization. Geochimi. Cosmo. Acta *75*, 3568–3580.

McDonald, J.H., and Kreitman, M. (1991). Adaptive protein evolution at the Adh locus in *Drosophila.* Nature *351*, 652–654.

McVean, G., Awadalla, P., and Fearnhead, P. (2002). A coalescent-based method for detecting and estimating recombination from gene sequences. Genetics *160*, 1231–1241.

Man, D., Wang, W., Sabehi, G., Aravind, L., Post, A.F., Massana, R., Spudich, E.N., Spudich, J.L., and Béjà, O. (2003). Diversification and spectral tuning in marine proteorhodopsins. EMBO J. *22*, 1725–1731.

Mann, N.H., Cook, A., Millard, A., Bailey, S., and Clokie, M. (2003). Marine ecosystems: Bacterial photosynthesis genes in a virus. Nature *424*, 741–741.

Martinsohn, J.T., Radman, M., and Petit, M.-A. (2008). The λ Red proteins promote efficient recombination between diverged sequences: implications for bacteriophage genome mosaicism. PLoS Genetics *4*, e1000065.

Mayr, E. (1942). Systematics and the Origin of Species – From the Viewpoint of a Zoologist (Havard University Press, Cambridge, MA).

Medini, D., Donati, C., Tettelin, H., Masignani, V., and Rappuoli, R. (2005). The microbial pan-genome. Curr. Opin. Gen. Dev. *15*, 589–594.

Mes, T.H. (2008). Microbial diversity – insights from population genetics. Environ. Microbiol. *10*, 251–264.

Minin, V.N., Dorman, K.S., Fang, F., and Suchard, M.A. (2005). Dual multiple change-point model leads to more accurate recombination detection. Bioinformatics *21*, 3034–3042.

Mueller, R.S., Denef, V.J., Kalnejais, L.H., Suttle, K.B., Thomas, B.C., Wilmes, P., Smith, R.L., Nordstrom, D.K., McCleskey, R.B., Shah, M.B., *et al.* (2010). Ecological distribution and population physiology defined by proteomics in a natural microbial community. Mol. Sys. Biol. *6*, 374.

Mueller, R.S., Dill, B.D., Pan, C., Belnap, C.P., Thomas, B.C., VerBerkmoes, N.C., Hettich, R.L., and Banfield, J.F. (2011). Proteome changes in the initial bacterial colonist during ecological succession in an acid mine drainage biofilm community. Environ. Microbiol. *13*, 2279–2292.

Pena, A., Teeling, H., Huerta-Cepas, J., Santos, F., Yarza, P., Brito-Echeverria, J., Lucio, M., Schmitt-Kopplin, P., Meseguer, I., Schenowitz, C., *et al.* (2010). Fine-scale evolution: genomic, phenotypic and ecological differentiation in two coexisting Salinibacter ruber strains. ISME J. *4*, 882–895.

Pérez-Losada, M., Browne, E.B., Madsen, A., Wirth, T., Viscidi, R.P., and Crandall, K.A. (2006). Population genetics of microbial pathogens estimated from multilocus sequence typing (MLST) data. Infect. Genet. Evol. *6*, 97–112.

Pernthaler, A., Dekas, A.E., Brown, C.T., Goffredi, S.K., Embaye, T., and Orphan, V.J. (2008). Diverse syntrophic partnerships from deep-sea methane vents revealed by direct cell capture and metagenomics. Proc. Natl. Acad. Sci. U.S.A. *105*, 7052–7057.

Petersen, L., Bollback, J.P., Dimmic, M., Hubisz, M., and Nielsen, R. (2007). Genes under positive selection in *Escherichia coli*. Genome Res. *17*, 1336–1343.

Piganeau, G., and Moreau, H. (2007). Screening the Sargasso Sea metagenome for data to investigate genome evolution in *Ostreococcus* (Prasinophyceae, Chlorophyta). Gene *406*, 184–190.

de Queiroz, K. (2005). Ernst Mayr and the modern concept of species. Proc. Natl. Acad. Sci. U.S.A. *102*, 6600–6607.

Ripp, S., Ogunseitan, O.A., and Miller, R.V. (1994). Transduction of a freshwater microbial community by a new *Pseudomonas aeruginosa* generalized transducing phage, UT1. Microb. Ecol. *3*, 121–126.

Roesch, L.F.W., Fulthorpe, R.R., Riva, A., Casella, G., Hadwin, A.K.M., Kent, A.D., Daroub, S.H., Camargo, F.A.O., Farmerie, W.G., and Triplett, E.W. (2007). Pyrosequencing enumerates and contrasts soil microbial diversity. ISME J *1*, 283–290.

Rossello-Mora, R., Lucio, M., Pena, A., Brito-Echeverria, J., Lopez-Lopez, A., Valens-Vadell, M., Frommberger, M., Anton, J., and Schmitt-Kopplin, P. (2008). Metabolic evidence for biogeographic isolation of the extremophilic bacterium *Salinibacter ruber*. ISME J. *2*, 242–253.

Rusch, D.B., Halpern, A.L., Sutton, G., Heidelberg, K.B., Williamson, S., Yooseph, S., Wu, D., Eisen, J.A., Hoffman, J.M., Remington, K., *et al.* (2007). The Sorcerer II Global Ocean Sampling Expedition: Northwest Atlantic through Eastern Tropical Pacific. PLoS Biol. *5*, e77.

Sherwood, C. (2003). Prophages and bacterial genomics: what have we learned so far? Mol. Microbiol. *49*, 277–300.

Shi, Y., Tyson, G.W., and DeLong, E.F. (2009). Metatranscriptomics reveals unique microbial small RNAs in the ocean's water column. Nature *459*, 266–269.

Simmons, S.L., DiBartolo, G., Denef, V.J., Goltsman, D.S.A., Thelen, M.P., and Banfield, J.F. (2008). Population genomic analysis of strain variation in *Leptospirillum* group II bacteria involved in acid mine drainage formation. PLoS Biol. *6*, e177.

Singer, S.W., Chan, C.S., Zemla, A., VerBerkmoes, N.C., Hwang, M., Hettich, R.L., Banfield, J.F., and Thelen, M.P. (2008). Characterization of Cytochrome 579, an unusual cytochrome isolated from an iron-oxidizing microbial community. Appl. Environ. Microbiol. *74*, 4454–4462.

Singer, S.W., Erickson, B.K., VerBerkmoes, N.C., Hwang, M., Shah, M.B., Hettich, R.L., Banfield, J.F., and Thelen, M.P. (2010). Posttranslational modification and sequence variation of redox-active proteins correlate with biofilm life cycle in natural microbial communities. ISME J. *4*, 1398–1409.

Sogin, M.L., Morrison, H.G., Huber, J.A., Welch, D.M., Huse, S.M., Neal, P.R., Arrieta, J.M., and Herndl, G.J. (2006). Microbial diversity in the deep sea and the underexplored "rare biosphere". Proc. Natl. Acad. Sci. U.S.A. *103*, 12115–12120.

Staley, J.T. (2009a). The *phylogenomic* species concept. Microbiol. Today *May 2009*, 80–83.
Staley, J.T. (2009b). The *phylogenomic* species concept for bacteria and archaea. Microbe 4, 361–365.
Thomas, C.M., and Nielsen, K.M. (2005). Mechanisms of, and barriers to, horizontal gene transfer between bacteria. Nat. Rev. Micro. 3, 711–721.
Thompson, J.R., Pacocha, S., Pharino, C., Klepac-Ceraj, V., Hunt, D.E., Benoit, J., Sarma-Rupavtarm, R., Distel, D.L., and Polz, M.F. (2005). Genotypic diversity within a natural coastal bacterioplankton population. Science *307*, 1311–1313.
Tringe, S.G., Zhang, T., Liu, X., Yu, Y., Lee, W.H., Yap, J., Yao, F., Suan, S.T., Ing, S.K., Haynes, M., *et al.* (2008). The airborne metagenome in an indoor urban environment. PLoS ONE 3, e1862.
Tyson, G.W., Chapman, J., Hugenholtz, P., Allen, E.E., Ram, R.J., Richardson, P.M., Solovyev, V.V., Rubin, E.M., Rokhsar, D.S., and Banfield, J.F. (2004). Community structure and metabolism through reconstruction of microbial genomes from the environment. Nature *428*, 37–43.
Venter, J.C., Remington, K., Heidelberg, J.F., Halpern, A.L., Rusch, D., Eisen, J.A., Wu, D., Paulsen, I., Nelson, K.E., Nelson, W., *et al.* (2004). Environmental genome shotgun sequencing of the Sargasso Sea. Science *304*, 66–74.
VerBerkmoes, N.C., Denef, V.J., Hettich, R.L., and Banfield, J.F. (2009). Systems biology: functional analysis of natural microbial consortia using community proteomics. Nat. Rev. Micro. 7, 196–205.
de Visser, J.A.G.M., and Rozen, D.E. (2006). Clonal interference and the periodic selection of new beneficial mutations in *Escherichia coli*. Genetics *172*, 2093–2100.
Vos, M. (2011). A species concept for bacteria based on adaptive divergence. Trends Microbiol. *19*, 1–7.
Wang, L., Spira, B., Zhou, Z., Feng, L., Maharjan, R.P., Li, X., Li, F., McKenzie, C., Reeves, P.R., and Ferenci, T. (2010). Divergence involving global regulatory gene mutations in an *Escherichia coli* population evolving under phosphate limitation. Genome Biol. Evol. *2*, 478–487.
Ward, D.M., Cohan, F.M., Bhaya, D., Heidelberg, J.F., Kuhl, M., and Grossman, A. (2007). Genomics, environmental genomics and the issue of microbial species. Heredity *2*, 207–219.
Ward, N. (2006). New directions and interactions in metagenomics research. FEMS Microbiol. Ecol. *55*, 331–338.
Wayne, L.G., Brenner, D.J., Colwell, R.R., Grimont, P.A.D., Kandler, O., Krichevsky, M.I., Moore, L.H., Moore, W.E.C., Murray, R.G.E., Stackebrandt, E., *et al.* (1987). Report of the *ad hoc* committee on reconciliation of approaches to bacterial systematics. Int. J System. Bacteriol. *37*, 463–464.
Wexler, M., Richardson, D.J., and Bond, P. (2009). Radiolabelled proteomics to determine differential functioning of Accumulibacter during the anaerobic and aerobic phases of a bioreactor operating for enhanced biological phosphorus removal. Environ. Microbiol. *11*, 3029–3044.
Whitaker, R.J., and Banfield, J.F. (2006). Population genomics in natural microbial communities. Trends Ecol. Evol. *21*, 508–516.
Wilhelm, L.J., Tripp, H.J., Givan, S.A., Smith, D.P., and Giovannoni, S.J. (2007). Natural variation in SAR11 marine bacterioplankton genomes inferred from metagenomic data. Biol. Direct *2*, 27.
Wilmes, P., and Bond, P.L. (2004). The application of two-dimensional polyacrylamide gel electrophoresis and downstream analyses to a mixed community of prokaryotic micro-organisms. Environ. Microbiol. *6*, 911–920.
Wilmes, P., and Bond, P.L. (2006). Towards exposure of elusive metabolic mixed-culture processes: the application of metaproteomic analyses to activated sludge. Water Sci. Tech. *54*, 217–226.
Wilmes, P., and Bond, P.L. (2009). Microbial community proteomics: elucidating the catalysts and metabolic mechanisms that drive the Earth's biogeochemical cycles. Curr. Opin. Microbiol. *12*, 310–317.
Wilmes, P., Andersson, A.F., Lefsrud, M.G., Wexler, M., Shah, M., Zhang, B., Hettich, R.L., Bond, P.L., VerBerkmoes, N.C., and Banfield, J.F. (2008a). Community proteogenomics highlights microbial strain-variant protein expression within activated sludge performing enhanced biological phosphorus removal. ISME J. *2*, 853–864.
Wilmes, P., Wexler, M., and Bond, P.L. (2008b). Metaproteomics provides functional insight into activated sludge wastewater treatment. PLoS ONE *3*, e1778.
Wilmes, P., Remis, J.P., Hwang, M., Auer, M., Thelen, M.P., and Banfield, J.F. (2009a). Natural acidophilic biofilm communities reflect distinct organismal and functional organization. ISME J. *3*, 266–270.
Wilmes, P., Simmons, S.L., Denef, V.J., and Banfield, J.F. (2009b). The dynamic genetic repertoire of microbial communities. FEMS Microbiol. Rev. *33*, 109–132.
Wilmes, P., Bowen, B.P., Thomas, B.C., Mueller, R.S., Denef, V.J., VerBerkmoes, N.C., Hettich, R.L., Northen, T.R., and Banfield, J.F. (2010). Metabolome-proteome differentiation coupled to microbial divergence. mBio *1*, e00246–00210.

Wilson, G.G., and Murray, N.E. (1991). Restriction and modification systems. Ann. Rev. Gen. 25, 585–627.

Woyke, T., Teeling, H., Ivanova, N.N., Huntemann, M., Richter, M., Gloeckner, F.O., Boffelli, D., Anderson, I.J., Barry, K.W., Shapiro, H.J., *et al.* (2006). Symbiosis insights through metagenomic analysis of a microbial consortium. Nature *443*, 950–955.

Zeidner, G., Bielawski, J.P., Shmoish, M., Scanlan, D.J., Sabehi, G., and Béjà, O. (2005). Potential photosynthesis gene recombination between *Prochlorococcus* and *Synechococcus* via viral intermediates. Environ. Microbiol. *7*, 1505–1513.

The Human–Microbe Coevolutionary Continuum

2

Lesley A. Ogilvie, Andrew D.J. Overall and Brian V. Jones

Abstract

Humans enter into a range of symbioses with resident and transiently colonizing microbes, which span a dynamic continuum from antagonistic to mutualistic. These interactions are shaped by a complex set of selective forces, which include both host and microbially derived selective pressures. Given the significant impact that both resident and pathogenic microbes can have on our health, there is now a move to develop a theoretical framework that may guide studies of human–microbe interactions. This should enable the deeper level of understanding required to model, predict and ultimately control human diseases related to antagonistic or aberrant host–microbe interactions. Here we explore the human–microbe coevolutionary continuum in the context of current and emerging theory, and with a focus on the opposite ends of the spectrum: mutualism and antagonism. In doing so we highlight areas in which theory is helping to enhance the understanding of this dynamic continuum and where current theory fails as well as suggesting future avenues of research.

Defining and detecting coevolution

Since Darwin emphasized the importance of the 'mutual interactions of organisms' and the possibility of coevolution occurring between bees and clover in *On the Origin of Species* (Darwin, 1859), coevolutionary theory has been a subject of intense speculation and investigation. Darwin regarded biotic interactions spanning the continuum from mutualistic to antagonistic as the core drivers of natural selection, creating complex species-rich ecosystems built upon coevolutionary processes that enable survival, growth and reproduction.

Primordial coevolutionary interactions between prokaryotes and eukaryotes have had a profound influence on the evolution of eukaryotic life forms. The existence of prokaryotes nearly 2 billion years before the emergence of eukaryotic life, in tandem with their ubiquitous distribution across almost every aquatic and terrestrial niche available, has led to the notion that prokaryotes have played a fundamental role in the evolution of eukaryotic life. The endogenization of primordial prokaryotes is widely accepted to have resulted in the development of mitochondria in eukaryotic cells, opening up evolutionary possibilities by conferring the ability to use oxygen to convert glucose into ATP, a storable form of energy (see Cavalier-Smith, 2009). Likewise, photosynthetic bacteria were the precursors of chloroplasts, which enable plant cells to synthesize glucose using light energy, and are believed to have formed by an analogous process to mitochondria (see Archibald, 2009).

A joint evolutionary trajectory between animals and microbes is also evident in the many examples of their fundamental roles in organismal development (McFall-Ngai and Ruby, 1991; McFall-Ngai, 2001, 2002; Ryan, 2009; Gilbert *et al.*, 2010). Notable examples include the development of specialized tissues within the sepiolid squid, *Euprymna scolopes*, the sole function of which is to enable colonization of the squid by its bioluminescent bacterial symbiont, *Vibrio fischeri* (McFall-Ngai and Ruby, 1991); the critical role of viral proteins in the development of the mammalian placenta and skin and eye structure, as well as perhaps more controversially their proposed roles in cancer (Ryan, 2009). Such examples not only illustrate the impact of microbes on development of higher eukaryotes, but also the potential for development of coevolutionary relationships between host and microbe, in which the evolution of certain traits in each participant is linked to the other.

Coevolution is defined here as the reciprocal change in genetic composition in response to the interaction between populations. Importantly for host–microbe coevolution, this can occur in any type of interaction from pathogenic to mutualistic. The coevolution of entities, such as species, does not preclude the influence of genetic drift on neutral coevolution: correlations that arise as a consequence of population size. Adaptive coevolution is distinguished from neutral coevolution by correlating phenotypic traits such as resistance and infectivity (in the host–pathogen model) or complementary/reciprocal traits in mutualistic interactions, as well as identifying the corresponding molecular footprint within the genomes of interacting species. Signals of positive selection can be identified within these genomes at the level of the gene if, for example, the nucleotides in codons that determine the particular amino acid incorporated into the final protein (non-synonymous sites; d*N*) are assumed to be less variable relative to those that do not (synonymous sites d*S*, or silent mutations) (see Jensen *et al.*, 2007); when $dN/dS > 1$, positive (or diversifying) selection is thought to be occurring, e.g. selection exerted by a pathogen, whereas if $dN/dS < 1$, negative (or purifying) selection is thought to be occurring. The extensive variations and developments within these approaches are reviewed comprehensively elsewhere (Aguileta *et al.*, 2009).

A human–microbe coevolutionary continuum

Humans enter into a range of symbioses with microbes spanning a dynamic continuum from antagonistic to mutualistic. These relationships are shaped by a complex set of selective pressures, which vary depending on the nature of the interaction and the type of association between microbe and host. However, whether the interaction is deleterious or beneficial, with primary pathogen or specialized mutualist and long term or intermittent evidence is accumulating that humans and other animals are focal points of a host–microbe coevolutionary continuum (Fig. 2.1).

At one end of this evolutionary continuum are mutualistic or beneficial relationships; with particular examples deriving from our interactions with the plethora of microbes normally found colonizing the human body. Each human being is home to a vast array of bacteria, archaea, fungi, viruses and protozoa. These microbes colonize every exposed surface of our bodies and a range of niches within, such as the respiratory tract, vagina and gastrointestinal tract (GIT). The majority of our microbial associates are bacteria, with every individual estimated to harbour around 10× more bacterial cells than human cells (Savage, 1977). The densest populations are found within the GIT where approximately

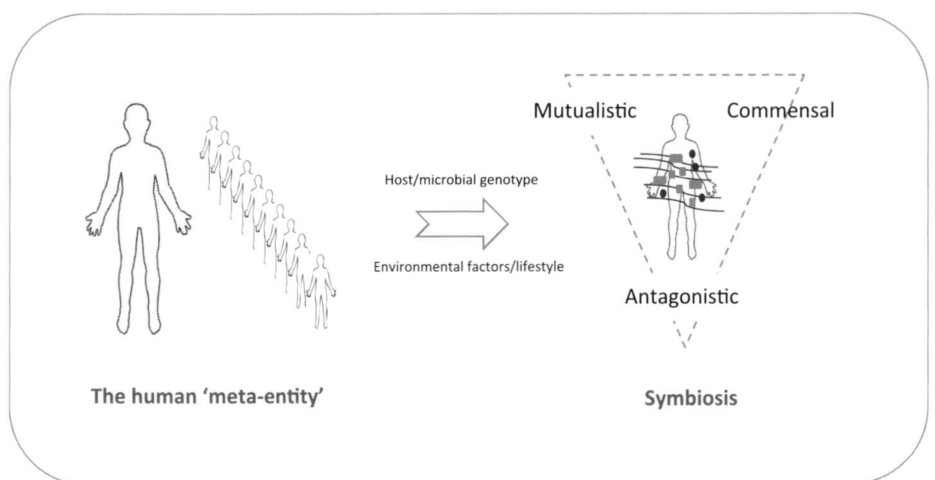

Figure 2.1 The human–microbe coevolutionary continuum. Each human being is home to a diverse group of microbiota, with bacterial cells outnumbering human cells by an estimated 10×. Humans enter into a range of symbioses with these resident microbes spanning a dynamic continuum from antagonistic to mutualistic. The nature of the interaction is shaped by a complex set of selective forces, which include both host and microbially derived pressures.

10^{13}–10^{14} prokaryotic cells reside in a community dominated by bacteria (Whitman *et al.*, 1998; Ley *et al.*, 2006).

Collectively, our associated microbes are referred to as the human microbiome, and their genetic content likened to a 'second' human genome. This microbiome has been defined as 'the community of micro-organisms including prokaryotes, viruses and microbial eukaryotes that populate the human body.' (Nelson *et al.*, 2010), and is generally considered to constitute those microbes that are long term, and in many cases life long, human residents. Many (but not all) constituents of the human microbiome enter into commensal or mutualistic relationships with their hosts, but opportunistic pathogens are also represented in these communities, highlighting the potential for antagonistic interactions when the chance arises (Fig. 2.1).

The gut microbiome in particular has emerged as a key example of host–microbe interaction from the mutualistic end of the human coevolutionary continuum. This, as with other host-associated microbial communities, is subject to unique selective pressures, in that interactions and competition between constituent members operate alongside pressures exerted at the level of the human host (owing to effects on host fitness) (Fig. 2.1 and Ley *et al.*, 2006). A growing body of evidence is now demonstrating that our gut microbiomes play an intimate role in our development and wellbeing and confer a number of benefits to the human host. These include the salvage of energy from food, synthesis of vitamins, protection against pathogens via competitive exclusion, degradation of xenobiotic compounds and maturation of the immune system, (e.g. McNeil, 1984; Guarner and Malagelada, 2003; Bäckhed *et al.*, 2005; Mazmanian *et al.*, 2005; Gill *et al.*, 2006; Ley *et al.*, 2006; Kurokawa *et al.*, 2007; Hehemann *et al.*, 2010). Studies comparing germ-free mice lacking a gut microbiome to their conventionally raised counterparts have also provided evidence for the incorporation of gut microbes into the normal developmental programme of the host

(Hooper, 2004) with germ-free animals displaying distinct physiological and anatomical abnormalities during their development. Moreover, these responses to our microbial residents are evolutionarily conserved across a number of organismal types (Ley *et al.*, 2009). Recent work is also suggesting that normal gut microbial residents may even have the ability to affect behaviour and mood (Heijtz *et al.*, 2011), and have potentially contributed to the development of social behaviours and group living in our ancestors (Lombardo, 2007).

At the other end of the continuum, antagonistic interactions between the human host and microbes (often occurring with those microbes that are only transiently resident within the human microbiome) arise that, despite vast improvements in therapeutics, can cause infections leading to diseases such as AIDS, tuberculosis and diarrhoea, which result in nearly 10 million deaths per year worldwide (Lopez *et al.*, 2006). This significant impact on human health is thought to be due, in part, to the development of resistance or emergence of pathogen variants. Owing to these rapid alterations in the genetic composition of important human pathogens such as the human immunodeficiency virus (HIV) and Influenza A virus (Fitch *et al.* 1997; Bush *et al.* 1999a,b; Shih *et al.* 2007; Neumann *et al.* 2009), in tandem with the significant anthropogenic alterations of the environment affecting the re/emergence of pathogens (Epstein, 2001; Zell, 2004), incorporation of a theoretical backbone into studies of human–microbe interactions may help to provide the deeper level of understanding required to predict and ultimately control certain human diseases (Best *et al.*, 2009).

The gut microbiota may also provide examples where antagonist interactions may extend beyond the classic pathogen–host model, in which the coevolution of a specific pathogen with its host is considered. Most importantly in this context, perturbations in the structure and functioning of the gut community has been linked to a range of diseases, including metabolic, autoimmune and inflammatory disorders, diabetes, atherosclerosis, obesity and even cancer (Guarner and Malagelada, 2003; Kinross *et al.*, 2008; Lepage *et al.*, 2008; Sanz *et al.*, 2008; Cerf-Bensussan and Gaboriau-Routhiau, 2010; Chow *et al.*, 2010; Kang *et al.*, 2010; Kinross *et al.*, 2011). The apparent scope for negative impact of this community as a whole on human heath has led to the hypothesis that the entire community may be considered 'pathogenic' in some cases (Ley *et al.*, 2006).

In light of the explosion of studies accessing the genetic and metabolic potential of the human microbiome, and detailing the mechanism underlying our interactions with pathogenic microbes outside of this community (e.g. Eckburg *et al.*, 2005; Kurokawa *et al.*, 2007; Verberkmoes *et al.*, 2009; Qin *et al.*, 2010; Nam *et al.*, 2011; Zheng *et al.*, 2011) the foundations are being laid for reshaping the coevolutionary landscape (Ley *et al.*, 2006; Dethlefsen *et al.*, 2007; Xu *et al.*, 2007; Zilber-Rosenberg and Rosenberg, 2008; Ley *et al.*, 2009; Jones, 2010). In an effort to gain a more cohesive appreciation of the coevolutionary process, which depends on both host and microbial genotypes and pleiotropic interactions between constituent members, we explore the human–microbe coevolutionary continuum in the context of current and emerging theory and with a focus on the opposite ends of the spectrum – antagonism and mutualism.

Antagonistic Red Queens

In the context of the antagonistic interaction between host and pathogen, the pathogen is under strong selective pressure to evolve traits that allow it to evade the host's defences and complete its development/infection cycle, while the host is under selective pressure to evolve traits that allow it to resist invasion by the pathogen. This evolution of traits such as these has often been described as an 'arms race' (Dawkins and Krebs, 1979) in which there are accumulated improvements in both host and pathogen, albeit capped by biological limitations (Thompson, 2005a) and the pleiotropic nature of genes.

One of the most enduring evolutionary frameworks for understanding host–pathogen coevolution is the Red Queen hypothesis (Van Valen, 1973). Van Valen was seeking to reconcile the idea that an organism can become adapted to its environment, yet can still be subject to constant adaptation. This scenario is often conceptualized using Sewell Wright's adaptive landscape (Wright, 1932), where the coordinates of the horizontal plane represent the genotypic or phenotypic variants and the vertical plane the corresponding fitness. This hill-climbing vision of evolution, where the forces of natural selection drive populations of a given trait upwards towards a local adaptive peak, has proven to be a useful framework for explaining the dynamics of evolution; most straightforwardly when the landscape is determined by relatively static abiotic factors. With regards to modelling the Red Queen hypothesis, the landscape experienced by one species will reflect the ever-changing state of coevolving species.

Because of this, there is no peak of optimal adaptation and the species is faced with the prospect of either evolving fast enough up a never-ending inclination or, otherwise, falling into a valley of extinction. The probability of extinction for any one species is, then, constant and independent of time; an observation that appears to be borne out in the paleontological record. These observations led Van Valen to the conclusion that coevolution will lead to the same probability of extinction in both participants owing to the exertion of reciprocal selection pressures driving a process of continuous evolution. If one participant fails to adapt then they may become extinct. Van Valen evoked the imagery of *Lewis Carroll's Through the Looking Glass* to illustrate these concepts (Carroll, 1871), in which the Red Queen tells Alice, 'Here, it takes all the running you can do, to keep in the same place'.

This hypothesis was extended to show, by simulation, that the selective pressures exerted by the host and parasite/pathogen would lead to long-term oscillations of genotype frequencies, and that these fluctuating dynamics could explain the existence and maintenance of sex and recombination as a means of maximizing genetic diversity within populations (Jaenike, 1978; Hamilton, 1980; Bell, 1985) and leading, potentially, to speciation (Buckling and Rainey, 2002; Brockhurst *et al.*, 2004). In the context of human hosts and pathogens, it is thought that although the reciprocal selective pressures exerted by hosts and their microbial pathogens will most likely result in oscillating allele frequencies in both populations, the selective pressures on human pathogens to evolve will not be strong enough until a large enough number of human hosts have the resistant phenotype, which apparently effectively slows the rate of pathogen evolution (Meyerson and Sawyer, 2011).

Despite its intuitive appeal, direct empirical confirmation of Red Queen dynamics in natural populations to date has been severely lacking. The only direct empirical demonstration comes from experimental evolution scenarios. These include infections with *Pseudomonas fluorescens* SBW25 and phage Φ2 (Paterson *et al.*, 2010) and on a multicellular level using the nematode *Caenorhabditis elegans* and the microbial parasite *Bacillus thuringiensis* (Schulte *et al.*, 2010). In the former study two evolutionary scenarios were compared: one in which only the bacterial genotype remained constant; and one in which both bacteria and phage were allowed to evolve.

Unlike previous studies that have relied solely on correlating infectivity and resistance,

Paterson and co-workers also provided a genetic window into the coevolutionary process. Comparative mutational analysis revealed the rate of molecular evolution was greater in the coevolving phage populations than in populations infecting a constant host genotype. Coevolving phage populations exhibited a greater level of genetic diversity, which was reflected in the extended range of hosts phage were able to infect. In accordance, the most rapidly evolving genes were involved in host infection. Likewise, increases in virulence of *Bacillus thuringiensis* and resistance of its nematode host were also observed, which were associated with reciprocal genetic change and increases in the rate of evolution (Schulte *et al.*, 2010).

Despite this convincing demonstration of Red Queen dynamics in experimental scenarios, it is difficult to extrapolate results from such simple experimental systems to the extremely complex ecological inputs of a natural environment, including the human body, which undoubtedly shape and constrain the coevolutionary relationships developing. Most empirical studies to date have investigated the nature of the antagonism between host and pathogen in isolation rather than in a more realistic setting in which multi-faceted interactions would be the norm.

Moreover, these studies largely ignore the influence of abiotic factors, as have many empirical evaluations of this theory. However, there is growing awareness that abiotic factors play a key role in the coevolutionary relationship, reducing or heightening the intensity of reciprocal selection; i.e. spatial variations in host–pathogen defence and counter-defence can occur owing to differences in the physical environment, creating hotspots and coldspots of coevolution – the 'geographic mosaic theory of coevolution' (Thompson, 1994, 2005b).

Concordantly, geographic variation in disease susceptibilities are well known; for example, sickle cell polymorphisms are maintained in areas where malaria is endemic and studies of suspected pathogen-mediated selection on human genes have uncovered an intriguing geographic variation in susceptibilities to infections; for instance, the CCR5-Δ32 deletion allele of the chemokine receptor 5 (CCR5) is carried by approximately 10% of Europeans, whereas it is virtually absent elsewhere (Samson *et al.*, 1996). This geographic variation is thought to have been driven by the aetiological agent of the bubonic plague, *Yersinia pestis* (Stephens *et al.*, 1998), or by the smallpox virus (Galvani and Slatkin, 2003), with carriers of the CCR5-Δ32 deletion allele more resistant to these diseases.

Within natural populations Red Queen dynamics have been tested in a range of environments and with a number of host–parasite combinations including the water flea *Daphnia magna* and its microparasites (Decaestecker *et al.*, 2007; Ebert, 2008), *Drosophila*–parasitoid interactions (Dupas *et al.*, 2009), the freshwater snail (*Potamopyrgus antipodarum*) and its trematode parasite (Lively and Dybdahl, 2000; Dybdahl and Lively, 2007; King *et al.*, 2011), and in long-term studies of the wild European rabbit and its pathogen the myxoma virus (Fenner and Fantini, 1999); however, these studies have only shown the nature of the coevolutionary interaction indirectly, relying in general on correlation of traits rather than reciprocity of genetic change.

Most recently the relevance of experimental evolutionary studies to the natural environment was tested by analysing phage-bacterium coevolution, essentially rerunning the study of Paterson and co-workers, but within a more 'natural' setting, i.e. soil microcosms (Gomez and Buckling, 2011). Because phage encounter their hosts passively and constitute only a fraction of the diversity of a community, it was unclear whether phage could exert a significant selective force on their bacterial hosts within a complex environment such as soil

(Gomez and Buckling, 2011). Gomez and Buckling were able to provide some evidence that coevolution was occurring within a natural environment, showing reciprocal adaptation of bacterium and phage, however, the dynamics of coevolution were altered in comparison with the simple experimental scenario reported by Paterson and co-workers (2010).

Rather than the 'arms race' (increasing levels of resistance over time), found in simple experimental conditions, fluctuating dynamics were observed, i.e. the intensity and direction of selection varied over time (see Bell, 2010). Bacteria were actually more resistant to contemporary than past and future phage and conversely phage could infect bacteria from the past and future better than bacteria from the present. Moreover, the high levels of resistance reported by Paterson and co-workers (2010) were not recapitulated in soil, presumably constrained by fitness costs imposed within the nutrient limited soil environment. Given the key role bacteria play within environments such as the human GIT, in tandem with the clinical significance of *Pseudomonas* spp. (Driscoll *et al.*, 2007) and thus the potential therapeutic use of phage, further elucidation of the coevolutionary dynamics of bacterial-phage interactions within natural environments is of particular significance.

Searching for the human–microbe coevolutionary imprint

When it comes to searching for evidence of host–pathogen coevolution within humans, the task is complicated by the long generation times and the complex nature of the interactions occurring. *In vitro* studies are providing some evidence that Red Queen dynamics are appropriate to describe host–pathogen interactions within the human body. A catalogue of studies on primate–virus interactions provides some convincing examples of Red Queen dynamics e.g. (Sawyer *et al.*, 2004, 2005; Zhang and Webb, 2004; Aragonès *et al.*, 2008; Ortiz *et al.*, 2009; Liu *et al.*, 2010); for example, comparison of non-synonymous versus synonymous substitutions showed that the hepatitis A virus readapted its codon usage in response to changing environmental cellular conditions in order to maintain its original level of virulence (Aragonès *et al.*, 2008).

In another example of pathogen-mediated selection, in this case HIV, molecular evolution studies of the interferon-inducible transmembrane protein tetherin and its antagonist virion infectivity factor (vif) have shown these molecules to be under positive selection by the viral protagonist. Other studies have used selection models to pin-point human genes with evidence of pathogen-mediated selection, including erythrocyte genes such as beta-globin, and the Duffy locus (Hamblin and Di Rienzo, 2000; Hamblin *et al.*, 2002), the TLR1 family (Huang *et al.*, 2011) and the major histocompatibility complex (MHC) (for in-depth review see Vallender and Lahn, 2004).

The MHC, a major component of the adaptive immune response of mammals and many other vertebrates, exhibits extremely high levels of diversity – up to 500 different alleles per gene locus – that has been linked to a range of infectious diseases (Haldane, 1949), including malaria, HIV and tuberculosis (Vallender and Lahn, 2004). Although MHC diversity may be driven by other factors such as gene conversion and recombination, MHC-dependent mate selection and maternal–fetal incompatibility (Edwards and Hedrick, 1998; Jeffery and Bangham, 2000; Nikolich-Zugich *et al.*, 2004); pathogen-mediated selection has been posited as a major selective force owing to the observation of higher levels of diversity at loci involved in pathogen recognition and host defence, than those involved in other functions.

Many of the existing examples of reciprocal interactions between two species, however, provide evidence for the classic *gene-for gene* model, rather than for the polygenic adaptations that will most likely occur in the human–microbe interaction. Further insight is promised by the recent leap in next-generation sequencing technology which is conferring a broader appreciation of the coevolutionary landscape, allowing detection of the molecular footprint of pathogen-mediated selection on a genome-wide basis (Sabeti *et al.*, 2006; Hancock and Di Rienzo, 2008; Enard *et al.*, 2010; Oleksyk *et al.*, 2010).

Predicting pathogen evolution

Although a plethora of studies are providing strong evidence for the existence of Red Queen dynamics with the human host–pathogen interaction, a key question remaining to be answered is whether these observations can be used to predict human host–pathogen evolution and more specifically the emergence of resistant pathogen phenotypes. Attempts have been made at predicting pathogen evolution such as estimating mutation rates in positively selected codons (Bush *et al.*, 1999a; Lee and Chen, 2004) or modelling the evolution of proteins in the presence of a selective pressure to infer coevolution between antigens and the immune system (Rohrbach and Dickerson, 2009). A cyclical arms race model has also been proposed in which adaption of a virus to a particular host genotype is constrained owing to the myriad interactions between genetic determinants. This limits the number of alleles that can be modified and thus these are recycled over time in a 'rock–paper–scissors' chase (Meyerson and Sawyer, 2011). Importantly, these recent studies are recognizing the complex web of interactions occurring between host and pathogen, pathogen and their microbial contemporaries and also at the molecular level, to provide the first (and credible) advances in the bid to harness the predictive powers of the theoretical framework.

In this respect recent work revealing the principal forces shaping horizontal gene transfer (HGT) (Smillie *et al.*, 2011), a major driving force of bacterial evolution (Ochman *et al.*, 2000; Koonin *et al.*, 2001) within the human microbiome, must be taken into account in any predictive scenario. Smillie and co-workers were able to show that gene transfer between bacteria subject to similar ecological forces was significantly enhanced in comparison with those of similar phylogenetic or geographical origin (Smillie *et al.*, 2011).

In a similar vein, further recent work has provided firm evidence of HGT between a human host and the obligate human pathogen *Neisseria gonorrhoeae*, showing a 689 bp fragment of the human long interspersed nuclear element L1 has been horizontally transferred to the pathogen (Anderson and Seifert, 2011). The function, if one exists, of this horizontally acquired DNA in *N. gonorrhoeae* is still a matter of debate (Anderson and Seifert, 2011; Shafer and Ohneck, 2011); however, the fact that human DNA has been horizontally transferred to a resident microbial pathogen (and conceivably our microbial commensals) has important consequences for human development and functioning and forces us to rethink our notions of human microbe coevolution. Such considerations should inform and underpin future theoretical models.

Mutualistic symbioses

In contrast to antagonistic interactions between host and microbial pathogen, the development and testing of ecological theories to explain mutualistic interactions within the human microbiome is still in its relative infancy. Theoretical studies of mutualistic relationships based on higher eukaryotes, have often taken a pairwise perspective ignoring the fact that most mutualistic interactions involve multispecies assemblages or guilds and may be part of a dynamic relationship that varies temporally and spatially from mutualistic to antagonistic (Gomulkiewicz *et al.*, 2003; Stanton, 2003); for example, the addition of just one species to a pairwise interaction was shown, theoretically, to change the nature of the interaction between two original species, provoking unidirectional changes or oscillations between mutualism and antagonism (depending on the prevailing conditions) (Gomulkiewicz *et al.*, 2003). This realization is especially pertinent in the human microbiome and the gut microbiota in particular, which presents a host-associated dynamic ecosystem subject to a range of 'top-down' and 'bottom-up' ecological and evolutionary forces, acting at the molecular to organismal level. This creates a highly competitive but also a highly cooperative community with many emergent properties (see Ley *et al.*, 2006).

In tandem another layer of complexity exists in the form of mobile genetic elements (MGE) and bacteriophages, collectively termed the mobile metagenome (Jones and Marchesi, 2007; Jones, 2010; Jones *et al.*, 2010). In general, the role of this mobile gene pool in the coevolution of host and microbe remains largely unexplored, but emerging evidence from studies of the human GIT, suggest this flexible gene pool plays a key role in the development of this community and host–microbe interactions (Claesson *et al.*, 2006; Corr *et al.*, 2007; Ebdon *et al.*, 2007; Kurokawa *et al.*, 2007; Li *et al.*, 2007; Xu *et al.*, 2007; Hehemann *et al.*, 2010; Jones, 2010; Jones *et al.*, 2010; Reyes *et al.*, 2010; Zaneveld *et al.*, 2010; Smillie *et al.*, 2011; Ogilvie *et al.*, 2012). Interestingly, MGE have the potential to facilitate cooperation between bacteria (Nogueira *et al.*, 2009; Rankin *et al.*, 2011) and therefore may be mechanisms by which microbial communities maintain cooperation and stability (for further discussion see also Ogilvie *et al.*, 2012). Also of particular relevance is the ability of MGEs to transfer genetic material across kingdom boundaries (Heinemann and Sprague, 1989; Shoemaker *et al.*, 2001; Goulas *et al.*, 2011), including transfer between bacterial and mammalian cells (Waters, 2001; Goulas *et al.*, 2011) as well as the ability of phage predation to shape the structure of microbial communities (Sandaa *et al.*, 2009; Shapiro *et al.*, 2010; Gomez and Buckling, 2011).

Symbioses within the human GIT

Before birth the human infant GIT is sterile, becoming colonized during delivery with microbiota from the mother and the surrounding environment. The infant GIT microbiota is highly variable, with mode of delivery (vaginal and caesarean), diet (breast of bottle-fed) and gestational age major factors affecting composition (Stark and Lee, 1982; Schwiertz *et al.*, 2003; Hällström *et al.*, 2004; Penders *et al.*, 2006). After around one year the infant microbiota generally resembles that of an adult, i.e. a highly individual but temporally stable phyla poor and species/strain rich community most often dominated by Bacteroidetes, Firmicutes, Actinobacteria, Proteobacteria and Verrucomicrobia (Zoetendal *et al.*, 1998; Eckburg *et al.*, 2005; Ley *et al.*, 2006) (Fig. 2.2). Phylogenies of the human GIT microbiota exhibit a unique fan-like structure thought to be suggestive of a coevolutionary history in which a limited number of early colonizers diversified into many different species and strains over a short evolutionary time frame (Bäckhed *et al.*, 2005; Ley *et al.*, 2006; Dethlefsen *et al.*, 2007; Chow *et al.*, 2010).

Understanding how coevolutionary forces act within a scenario in which pleiotropic effectors interact to develop and maintain symbiotic relationships as found in the human GIT, presents a great challenge. Traditional hypotheses on coevolving symbiotic mutualism predict that selection will favour reduced diversity to at least a small network of

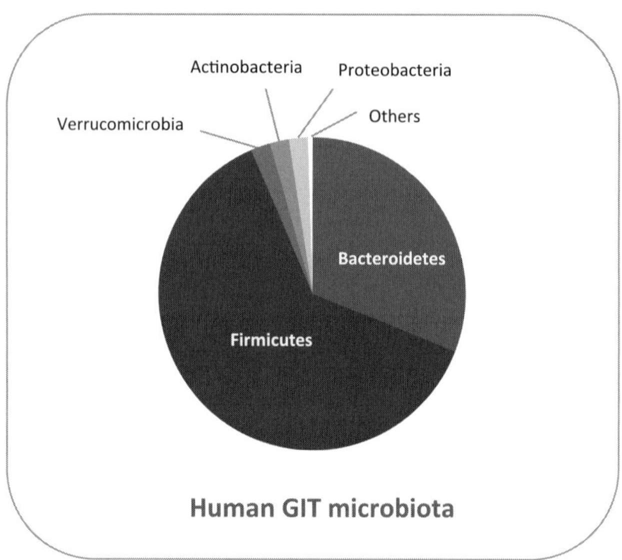

Figure 2.2 Major phylogenetic divisions of bacteria within the human gut microbiome. The human gastrointestinal tract is a phyla poor, but species rich environment, dominated by members of the Bacteroidetes and Firmicutes.

complementary symbiotic species; however, these are inadequate to describe the structure, functioning and persistence of large assemblages of symbiotic mutualists such as those found in the human GIT. Current theory suggests that organisms such as humans which harbour complex microbial communities can partition them spatially or according to available resources (niche partitioning) (Thompson, 2005b). Accordingly, resource partitioning has been shown to occur in environments where high levels of microdiversity (<1% divergence of 16S rRNA sequences) exist, including ocean surface water (Acinas *et al.*, 2004), acid mine drainage biofilm (Denef *et al.*, 2010), estuarine sediments (Oakley *et al.*, 2011) and has also been suggested to occur in the human gut (Eckburg *et al.*, 2005). Such high levels of strain-level genomic variation, as found within the human GIT, are thought to be a major determinant of distinct ecological trajectories within communities (Denef *et al.*, 2010).

It could be that the organization of microbial communities within the human GIT as biofilms may promote the existence of micro-scale niches which in turn promote microdiversity, as has been indicated for other microbial communities (Wilmes *et al.*, 2009; Ma and Banfield, 2011). In the human GIT, however, diversification of niches is seemingly hampered by the lack of energy sources, in comparison to other environments, with the GIT microbiota deriving energy from carbon (fermentation and methanogenesis), sulfate (sulfate reduction) and hydrogen only. Moreover, the constant mixing of gut contents reduces the capacity to spatially partition resources (for further discussion see Ley *et al.*, 2006). Further studies elucidating differences between the structure, functioning and distinct ecological niches of mucosal-associated biofilm communities and luminal communities are required to further our knowledge of mutualistic interactions within the GIT.

The hologenome theory of evolution

One of the most pervasive theories to emerge to describe the human–microbe relationship is the 'hologenome theory' of evolution proposed by Rosenberg and co-workers (Rosenberg *et al.*, 2007). The theory originates from observations on the symbiotic interaction between corals and their large resident populations of eukaryotic algae, bacteria and archaea which confer various benefits to their hosts (Rosenberg *et al.*, 2007), and in essence posits the idea of the holobiont, i.e. the complex metazoan and its associated microbial community, being the unit of selection in evolution (Rosenberg *et al.*, 2007, 2009; Zilber-Rosenberg and Rosenberg, 2008). The theory is based on four central tenets: (i) symbiotic relationships are established between microorganisms and all animals (and plants); (ii) there is transmission of symbiotic organisms between generations; (iii) the fitness of the holobiont within its environment is affected by the interactions occurring between hosts and symbionts; (iv) changes in host or associated microbial genomes will result in variation in the hologenome, i.e. the sum of the genetic information contained within the host and its microbiota. Such an arrangement will increase the ability of a human host to adapt to fluctuating or new environmental conditions. In this way, the microbial residents of complex metazoans play an essential and complementary role in the adaption and evolution of higher organisms (Zilber-Rosenberg and Rosenberg, 2008).

The hologenome theory is built upon a backdrop of theory and observation from macro-ecology, with the term 'holobiont' coined in referral to a collection of symbionts (Margulis and Fester, 1991), and the concept of the 'superorganism', defined as 'collection of single creatures that together possess of organism' (see Wilson and Sober, 1989). Indeed, Rosenberg and colleagues evoke group selection by introducing the notion that the organisms constituting the holobiont function and interact together – co-operate or compete – as a community so they can adapt to prevailing environmental conditions. Moreover, distinct Lamarkian aspects are inherent in the assumptions of the hologenome theory, in that environmental forces can lead to change in the amount and type of symbiotic microorganisms that are part of the holobiont and that these environmentally induced changes are heritable (Rosenberg *et al.*, 2009).

The flexible nature of the prokaryotic genome, with its ability to acquire new functions by HGT and short generation times provides a degree of flexibility to the relatively static human genomic complement (Rosenberg *et al.*, 2007; Zilber-Rosenberg and Rosenberg, 2008; Jones, 2010). This fits well with the growing appreciation of the significant role of the mobile metagenome in human fitness (Jones, 2010; Ogilvie *et al.*, 2012), in that the genetic information encoded by the prokaryotic complement of the holobiont can be regarded as a secondary accessory genome, whereas the mobile metagenome can now be added as a third, highly flexible, genetic pool which increases the ability of the holobiont to adapt to prevailing environmental conditions (Jones, 2010; Ogilvie *et al.*, 2012).

Current research by the Rosenberg group is now beginning to demonstrate the hologenome theory in action, suggesting that commensal bacteria can play a role in mate choice by the fruit fly, *Drosophila melanogaster* (Sharon *et al.*, 2010). They found that antibiotics were able to dismantle mating choices between groups of *D. melanogaster* populations reared on differential feeding regimes, suggesting and further supporting the idea that symbiotic bacteria have an integral role in the adaptation and evolution of higher organisms.

However, certain facets of the theory remain to be elucidated. One major point is the definition of the unit of selection: if the unit of selection is the entity whose survival and reproduction determines how the population of units evolves, could the level of the holobiont could break down in the presence of a single devastating mutant microbial species, or gene for that matter. Is this what we observe during infection and the onset and progression of disease? Modelling the evolutionary stable strategies of these scenarios remains to be explored, i.e. what it is about the holobiont that could evolve; what are the properties that could be measured? Nevertheless, this ongoing debate as to what constitutes a unit of selection should not detract from what the hologenome theory offers as a model of human–microbe coevolution. Although it remains to be seen how this relatively new theory will develop in this rapid pace of microbial discovery, we look forward to seeing how we can harness the predictive powers of the theoretical framework as the theory is developed.

Perspective

The dawn of the 'meta-omics' era has allowed us to apply high-resolution molecular techniques to environmental samples, enabling unparalleled exploration of community structure and function in human microbomes. This has permitted novel ecological and evolutionary insights that will potentially impact our comprehension of the development and progression of disease, and provide answers to fundamental ecological questions. Recent large-scale sequencing efforts have shown that bacterial residents of the human GIT comprise up to 1000 different species that collectively encode 100 times more genes than our own human cells (Gill *et al.*, 2006; Yang *et al.*, 2009; Qin *et al.*, 2010). Moreover, The International Human Genome Sequencing Consortium (Lander *et al.*, 2001) estimates that at least 1 in every 30 genes encodes functions with roles in defence and immunity, paying strong testament to the ongoing coevolutionary relationship we have with microbial pathogens and highlighting the significant evolutionary impact microbes have had (and are still having) on human development.

We are at an exciting junction in exploring and defining human–microbe interactions, progressing from describing 'who's there?' to now asking 'what are they doing?' and 'how can we use and protect them?'; fundamental ecological questions that have confounded (and still are) macro-ecologists for decades.

Current data are revealing the complexity of the task ahead and highlighting the necessity to view the relationships between the human host, its microbiome and its transient pathogenic visitors as dynamic continuum spanning from antagonistic to mutualistic. Recent research is also recognizing the myriad of interactions occurring between host and microbe (pathogens and commensals), between microbes and their contemporaries and also at the molecular level. Moreover, a number of areas requiring further investigation have been highlighted, including the role of MGE in the coevolutionary process and the use of whole genome sequencing to provide credible insight into the coevolutionary process. These future insights will inform and underpin human–microbe coevolutionary theories developed.

Macro-ecological theory is providing a solid base for the exploration of the human–microbe coevolutionary continuum and can be regarded as a starting point for the development of new insight and perspective. Exploration of the human–microbe coevolutionary landscape is at an early stage, however, nascent theories and perspectives, some explored here, are beginning to reshape our understanding of the association of human beings and microbes, with the potential to impact our ability to prevent, prediction and cure human diseases.

Acknowledgement

Lesley A. Ogilvie is supported by a Medical Research Council funded fellowship awarded to Brian V. Jones (Grant ID number 93344/G0901553), and receives funding from the University of Brighton.

References

Acinas, S.G., Klepac-Ceraj, V., Hunt, D.E., Pharino, C., Ceraj, I., Distel, D.L., and Polz, M.F. (2004). Fine-scale phylogenetic architecture of a complex bacterial community. Nature *430*, 551–554.

Aguileta, G., Refrégier, G., Yockteng, R., Fournier, E., and Giraud, T. (2009). Rapidly evolving genes in pathogens: methods for detecting positive selection and examples among fungi, bacteria, viruses and protists. Infect. Genet. Evol. *9*, 656–670.

Anderson, M.T., and Seifert, H.S. (2011). Opportunity and means: horizontal gene transfer from the human host to a bacterial pathogen. mBio *2*, e00005–e00011.

Aragonès, L., Bosch, A., and Pintó, R.M. (2008). Hepatitis A virus mutant spectra under the selective pressure of monoclonal antibodies: codon usage constraints limit capsid variability. J. Virol. *82*, 1688–1700.

Archibald, J.M. (2009). The puzzle of plastid evolution. Curr. Biol. *19*, R81–R88.

Bäckhed, F., Ley, R.E., Sonnenburg, J.L., Peterson, D.A., and Gordon, J.I. (2005). Host-bacterial mutualism in the human intestine. Science *307*, 1915–1920.

Bell, G. (1985). Two theories of sex and variation. Experentia *41*, 1235–1245.

Bell, G. (2010). Fluctuating selection: the perpetual renewal of adaptation in variable environments. Phil. Trans. Roy. Soc. B*365*, 87–97.

Best, A., White, A., and Boots, M. (2009). The implications of coevolutionary dynamics to host–parasite interactions. Am. Nat. *173*, 779–791.

Brockhurst, M.A., Rainey, P.B., and Buckling, A. (2004). The effect of spatial heterogeneity and parasites on the evolution of host diversity. Phil. Trans. Roy. Soc. B *271*, 107–111.

Buckling, A., and Rainey, P.B. (2002). The role of parasites in sympatric and allopatric host diversification. Nature *420*, 496–499.

Bush, R., Bender, C., Subbarao, K., Cox, N., and Fitch, W. (1999a). Predicting the evolution of human influenza A. Science *286*, 1921–1925.

Bush, R., Fitch, W., Bender, C., and Cox, N. (1999b). Positive selection on the H3 hemagglutinin gene of human influenza virus A. Mol. Biol. Evol. *16*, 1457–1465.

Carroll, L. (1871). Through the Looking-Glass, and What Alice Found There (Penguin Books, London).

Cavalier-Smith, T. (2009). Predation and eukaryote cell origins: a coevolutionary perspective. Int. J. Biochem. Cell Biol. *41*, 307–322.

Cerf-Bensussan, N., and Gaboriau-Routhiau, V. (2010). The immune system and the gut microbiota: friends or foes? Nat. Rev. Immun. *10*, 735–744.

Chow, J., Lee, S.M., Shen, Y., Khosravi, A., and Mazmanian, S.K. (2010). Host–bacterial symbiosis in health and disease. Adv. Immunol. *107*, 243–274.

Claesson, M.J., Li, Y., Leahy, S., Canchaya, C., van Pijkeren, J.P., Cerdeño-Tárraga, A.M., Parkhill, J., Flynn, S., O'Sullivan, G.C., Collins, J.K., *et al.* (2006). Multireplicon genome architecture of Lactobacillus salivarius. Proc. Natl. Acad. Sci. U.S.A. *103*, 6718–6723.

Corr, S.C., Li, Y., Riedel, C.U., O'Toole, P.W., Hill, C., and Gahan, C.G. (2007). Bacteriocin production as a mechanism for the antiinfective activity of Lactobacillus salivarius UCC118. Proc. Natl. Acad. Sci. U.S.A. *104*, 7617–7621.

Darwin, C. (1859). On the Origin of Species (John Murray, London).

Dawkins, R., and Krebs, J.R. (1979). Arms races between and within species. Phil. Trans. Roy. Soc. B *205*, 489–511.

Decaestecker, E., Gaba, S., Raeymaekers, J.A., Stoks, R., Van Kerckhoven, L., Ebert, D., and De Meester, L. (2007). Host-parasite 'Red Queen' dynamics archived in pond sediment. Nature *450*, 870–873.

Denef, V.J., Kalnejais, L.H., Mueller, R.S., Wilmes, P., Baker, B.J., Thomas, B.C., VerBerkmoes, N.C., Hettich, R.L., and Banfield, J.F. (2010). Proteogenomic basis for ecological divergence of closely related bacteria in natural acidophilic microbial communities. Proc. Natl. Acad. Sci. U.S.A. *107*, 2383–2390.

Dethlefsen, L., Mcfall-ngai, M., and Relman, D.A. (2007). An ecological and evolutionary perspective on human – microbe mutualism and disease. Nature *449*, 811–818.

Driscoll, J.A., Brody, S.L., and Kollef, M.H. (2007). The epidemiology, pathogenesis and treatment of Pseudomonas aeruginosa infections. Drugs *67*, 351–368.
Dupas, S., Dubuffet, A., Carton, Y., and Poirié, M. (2009). Local, geographic and phylogenetic scales of coevolution in Drosophila–parasitoid interactions. Adv. Parasitol. *70*, 281–295.
Dybdahl, M.F., and Lively, C.M. (2007). Host-Parasite coevolution: evidence for rare advantage and time-lagged selection in a natural population. Evolution 52, 1057–1067.
Ebdon, J., Muniesa, M., and Taylor, H. (2007). The application of a recently isolated strain of Bacteroides (GB-124) to identify human sources of faecal pollution in a temperate river catchment. Water Res. *41*, 3683–3690.
Ebert, D. (2008). Host–parasite coevolution: Insights from the Daphnia–parasite model system. Curr. Opin. Microbiol. *11*, 290–301.
Eckburg, P.B., Bik, E.M., Bernstein, C.N., Purdom, E., Dethlefsen, L., Sargent, M., Gill, S.R., Nelson, K.E., and Relman, D.A. (2005). Diversity of the human intestinal microbial flora. Science *308*, 1635–1638.
Edwards, S.V., and Hedrick, P.W. (1998). Evolution and ecology of MHC molecules: from genomics to sexual selection. Trends Ecol. Evol. *13*, 305–311.
Enard, D., Depaulis, F., and Roest Crollius, H. (2010). Human and non-human primate genomes share hotspots of positive selection. PLoS Gen. *6*, e1000840.
Epstein, P.R. (2001). Climate change and emerging infectious diseases. Microb. Infect. *3*, 747–754.
Fenner, F., and Fantini, B. (1999). Biological Control of Vertebrate Pests: The History of Myxomatosis – an Experiment in Evolution (CABI, Wallingford, UK).
Fitch, Bush, R., Bender, C., and Cox, N. (1997). Long term trends in the evolution of H(3) HA1 human influenza type A. Proc. Natl. Acad. Sci. U.S.A. *94*, 7712–7718.
Galvani, A.P., and Slatkin, M. (2003). Evaluating plague and smallpox as historical selective pressures for the CCR5-Delta 32 HIV-resistance allele. Proc. Natl. Acad. Sci. U.S.A. *100*, 15276–15279.
Gilbert, S.F., McDonald, E., Boyle, N., Buttino, N., Gyi, L., Mai, M., Prakash, N., and Robinson, J. (2010). Symbiosis as a source of selectable epigenetic variation: taking the heat for the big guy. Phil. Trans. Roy. Soc. B *365*, 671–678.
Gill, S.R., Pop, M., Deboy, R.T., Eckburg, P.B., Turnbaugh, P.J., Samuel, B.S., Gordon, J.I., Relman, D.A., Fraser-Liggett, C.M., and Nelson, K.E. (2006). Metagenomic analysis of the human distal gut microbiome. Science *312*, 1355–1359.
Gomez, P., and Buckling, A. (2011). Bacteria-phage antagonistic coevolution in soil. Science *332*, 106–109.
Gomulkiewicz, R., Nuismer, S.L., and Thompson, J.N. (2003). Coevolution in variable mutualisms. Am. Nat. *162*, S80–S93.
Goulas, T., Arolas, J.L., and Gomis-Rüth, F.X. (2011). Structure, function and latency regulation of a bacterial enterotoxin potentially derived from a mammalian adamalysin/ADAM xenolog. Proc. Natl. Acad. Sci. U.S.A. *108*, 1856–1861.
Guarner, F., and Malagelada, J.-R. (2003). Gut flora in health and disease. Lancet *361*, 512–519.
Haldane, J.B.S. (1949). Disease and evolution. Ric. Sci. *19*(Suppl. A), 68–76.
Hällström, M., Eerola, E., Vuento, R., Janas, M., and Tammela, O. (2004). Effects of mode of delivery and necrotising enterocolitis on the intestinal microflora in preterm infants. Eur. J. Clin. Microbiol. Infect. Dis. *23*, 463–470.
Hamblin, M., and Di Rienzo, A. (2000). Detection of the signature of natural selection in humans: evidence from the Duffy blood group locus. Am. J. Hum. Genet. *66*, 1669–1679.
Hamblin, M., Thompson, E., and Di Rienzo, A. (2002). Complex signatures of natural selection at the Duffy blood group locus. Am. J. Hum. Genet. *70*, 369–383.
Hamilton, W. (1980). Sex versus non-sex versus parasite. Oikos *35*, 282–290.
Hancock, A., and Di Rienzo, A. (2008). Detecting the genetic signature of natural selection in human populations: models, methods, and data. Ann. Rev. Anthropol. 37, 97–217.
Hehemann, J.-H., Correc, G., Barbeyron, T., Helbert, W., Czjzek, M., and Michel, G. (2010). Transfer of carbohydrate-active enzymes from marine bacteria to Japanese gut microbiota. Nature *464*, 908–912.
Heijtz, R.D., Wang, S., Anuar, F., Qian, Y., Björkholm, B., Samuelsson, A., Hibberd, M.L., Forssberg, H., and Pettersson, S. (2011). Normal gut microbiota modulates brain development and behavior. Proc. Natl. Acad. Sci. U.S.A. *108*, 3047–3052.
Heinemann, J.A., and Sprague, G.F. (1989). Bacterial conjugative plasmids mobilize DNA transfer between bacteria and yeast. Nature *340*, 205–209.
Hooper, L.V. (2004). Bacterial contributions to mammalian gut development. Trends Microbiol. *12*, 129–134.

Huang, Y., Temperley, N.D., Ren, L., Smith, J., Li, N., and Burt, D.W. (2011). Molecular evolution of the vertebrate TLR1 gene family – a complex history of gene duplication, gene conversion, positive selection and co-evolution. BMC Evo. Biol. *11*, 149.

Jaenike, J. (1978). An hypothesis to account for the maintenance of sex within populations. Evol. Theory *3*, 191–194.

Jeffery, K.J., and Bangham, C.R. (2000). Do infectious diseases drive MHC diversity? Microb. Infect. *2*, 1335–1341.

Jensen, J.D., Wong, A., and Aquadro, C.F. (2007). Approaches for identifying targets of positive selection. Trends Genet. *23*, 568–577.

Jones, B.V. (2010). The human gut mobile metagenome: a metazoan perspective. Gut Microbes *1*, 415–431.

Jones, B.V., and Marchesi, J.R. (2007). Transposon-aided capture (TRACA) of plasmids resident in the human gut mobile metagenome. Nat. Meth. *4*, 55–61.

Jones, B.V., Sun, F., and Marchesi, J.R. (2010). Comparative metagenomic analysis of plasmid encoded functions in the human gut microbiome. BMC Genomics *11*, 46.

Kang, S., Denman, S.E., Morrison, M., Yu, Z., Dore, J., Leclerc, M., and McSweeney, C.S. (2010). Dysbiosis of fecal microbiota in Crohn's disease patients as revealed by a custom phylogenetic microarray. Inflamm Bowel Dis. *16*, 2034–2042.

King, K.C., Jokela, J., and Lively, C.M. (2011). Parasites, sex, and clonal diversity in natural snail populations. Evolution *65*, 1474–1481.

Kinross, J.M., Darzi, A.W., and Nicholson, J.K. (2011). Gut microbiome–host interactions in health and disease. Genome Med. *3*, 14.

Kinross, J.M., von Roon, A.C., Holmes, E., Darzi, A., and Nicholson, J.K. (2008). The human gut microbiome: implications for future health care. Curr. Gastro. Rep. *10*, 396–403.

Koonin, E.V., Makarova, K.S., and Aravind, L. (2001). Horizontal gene transfer in prokaryotes: quantification and classification. Annu. Rev. Microbiol. *55*, 709–742.

Kurokawa, K., Itoh, T., Kuwahara, T., Oshima, K., Toh, H., Toyoda, A., Takami, H., Morita, H., Sharma, V.K., Srivastava, T.P., *et al.* (2007). Comparative metagenomics revealed commonly enriched gene sets in human gut microbiomes. DNA Res. *14*, 169–181.

Lander, E.S., Linton, L.M., Birren, B., Nusbaum, C., Zody, M.C., Baldwin, J., Devon, K., Dewar, K., Doyle, M., FitzHugh, W., *et al.* (2001). Initial sequencing and analysis of the human genome. Nature *409*, 860–921.

Lederberg, J. (2000). Infectious history. Science *288*, 287–293.

Lee, M.-S., and Chen, J.S.-E. (2004). Predicting antigenic variants of influenza A/H3N2 viruses. Emerg. Infect. Dis. *10*, 1385–1390.

Lepage, P., Colombet, J., Marteau, P., Sime-Ngando, T., Doré, J., and Leclerc, M. (2008). Dysbiosis in inflammatory bowel disease: a role for bacteriophages? Gut *57*, 424–425.

Ley, R.E., Peterson, D.A, and Gordon, J.I. (2006). Ecological and evolutionary forces shaping microbial diversity in the human intestine. Cell *124*, 837–848.

Ley, R.E., Lozupone, C.A., Hamady, M., Knight, R., and Jeffrey, I. (2008). Worlds within worlds : evolution of the vertebrate gut microbiota. Nat. Rev. Microbiol. *6*, 776–88.

Li, Y., Canchaya, C., Fang, F., Raftis, E., Ryan, K.A., van Pijkeren, J.-P., van Sinderen, D., and O'Toole, P.W. (2007). Distribution of megaplasmids in *Lactobacillus salivarius* and other lactobacilli. J. Bacteriol. *189*, 6128–6139.

Liu, J., Chen, K., Wang, J.-H., and Zhang, C. (2010). Molecular evolution of the primate antiviral restriction factor tetherin. PloS ONE *5*, e11904.

Lively, C.M., and Dybdahl, M.F. (2000). Parasite adaptation to locally common host genotypes. Nature *405*, 679–681.

Lombardo, M.P. (2007). Access to mutualistic endosymbiotic microbes: an underappreciated benefit of group living. Behav. Ecol. Sociobiol. *62*, 479–497.

Lopez, A., Mathers, C., Jamison, D., and Murray, C.J.L., eds. (2006). Global Burden of Disease and Risk Factors (Oxford University Press, New York).

Ma, S., and Banfield, J.F. (2011). Micron-scale Fe2+/Fe3+, intermediate sulfur species and O2 gradients across the biofilm–solution–sediment interface control biofilm organization. Geochim. Cosmochim. Ac. *75*, 3568–3580.

McFall-Ngai, M. (2001). Identifying 'prime suspects': symbioses and the evolution of multicellularity. Comp. Biochem. Physiol. B Biochem. Mol. Biol. *129*, 711–723.

McFall-Ngai, M. (2002). Unseen forces: the influence of bacteria on animal development. Dev. Biol. *242*, 1–14.

McFall-Ngai, M., and Ruby, E. (1991). Symbiont recognition and subsequent morphogenesis as early events in an animal–bacterial mutualism. Science *254*, 1491–1494.

McNeil, N.I. (1984). The contribution of the large intestine to energy supplies in man. Am. J. Clin. Nutr. *39*, 338–342.

Margulis, L., and Fester, R. (1991). Symbiosis as a Source of Evolutionary Innovation: Speciation and Morphogenesis (MIT Press, Cambridge, MA).

Mazmanian, S.K., Liu, C.H., Tzianabos, A.O., and Kasper, D.L. (2005). An immunomodulatory molecule of symbiotic bacteria directs maturation of the host immune system. Cell *122*, 107–118.

Meyerson, N.R., and Sawyer, S.L. (2011). Two-stepping through time: mammals and viruses. Trends Microbiol. *19*, 286–294.

Nam, Y.-D., Jung, M.-J., Roh, S.W., Kim, M.-S., and Bae, J.-W. (2011). Comparative analysis of Korean human gut microbiota by barcoded pyrosequencing. PLoS ONE *6*, e22109.

Neumann, G., Noda, T., and Kawaoka, Y. (2009). Emergence and pandemic potential of swine-origin H1N1 influenza virus. Nature *459*, 931–939.

Nikolich-Zugich, J., Fremont, D.H., Miley, M.J., and Messaoudi, I. (2004). The role of mhc polymorphism in anti-microbial resistance. Microb. Infect. *6*, 501–512.

Nogueira, T., Rankin, D.J., Touchon, M., Taddei, F., Brown, S.P., and Rocha, E.P. (2009) Horizontal gene transfer of the secretome drives the evolution of bacterial cooperation and virulence. Curr. Biol. *19*, 1683–1691.

Oakley, B.B., Carbonero, F., Dowd, S.E., Hawkins, R.J., and Purdy, K.J. (2011). Contrasting patterns of niche partitioning between two anaerobic terminal oxidizers of organic matter. ISME J. doi: 10.1038/ismej.2011.165. [Epub ahead of print]

Ochman, H., Lawrence, J.G., and Groisman, E.A. (2000). Lateral gene transfer and the nature of bacterial innovation. Nature *405*, 299–304.

Ogilvie, L.A., Firouzmand, S., and Jones, B.V. (2012). Evolutionary, ecological, and biotechnological perspectives on plasmids resident in the human gut mobile metagenome. Bioeng. Bugs *3*, 13–31.

Oleksyk, T.K., Smith, M.W., and O'Brien, S.J. (2010). Genome-wide scans for footprints of natural selection. Phil. Trans. Roy. Soc. B *365*, 185–205.

Ortiz, M., Guex, N., Patin, E., Martin, O., Xenarios, I., Ciuffi, A., Quintana-Murci, L., and Telenti, A. (2009). Evolutionary trajectories of primate genes involved in HIV pathogenesis. Molec. Biol. Evol. *26*, 2865–2875.

Paterson, S., Vogwill, T., Buckling, A., Benmayor, R., Spiers, A.J., Thomson, N.R., Quail, M., Smith, F., Walker, D., Libberton, B., *et al.* (2010). Antagonistic coevolution accelerates molecular evolution. Nature *464*, 275–278.

Penders, J., Thijs, C., Vink, C., Stelma, F.F., Snijders, B., Kummeling, I., van den Brandt, P.A., and Stobberingh, E.E. (2006). Factors influencing the composition of the intestinal microbiota in early infancy. Pediatrics *118*, 511–521.

Qin, J., Li, R., Raes, J., Arumugam, M., Burgdorf, K.S., Manichanh, C., Nielsen, T., Pons, N., Levenez, F., Yamada, T., *et al.* (2010). A human gut microbial gene catalogue established by metagenomic sequencing. Nature *464*, 59–65.

Rankin, D.J., Rocha, E.P., and Brown, S.P. (2011) What traits are carried on mobile genetic elements, and why? Heredity *106*,1–10.

Reyes, A., Haynes, M., Hanson, N., Angly, F.E., Heath, A.C., Rohwer, F., and Gordon, J.I. (2010). Viruses in the faecal microbiota of monozygotic twins and their mothers. Nature *466*, 334–338.

Rohrbach, A.S., and Dickerson, T.J. (2009). Predicting protein evolution *in vitro* by phage escape technology. Mol. Biosyst. *5*, 128–133.

Rosenberg, E., Koren, O., Reshef, L., Efrony, R., and Zilber-Rosenberg, I. (2007). The role of micro-organisms in coral health, disease and evolution. Nat. Rev. Microbiol. *5*, 355–362.

Rosenberg, E., Sharon, G., and Zilber-Rosenberg, I. (2009). The hologenome theory of evolution contains Lamarckian aspects within a Darwinian framework. Environ. Microbiol. *11*, 2959–2962.

Ryan, F. (2009). Virolution (Collins, London).

Sabeti, P.C., Schaffner, S.F., Fry, B., Lohmueller, J., Varilly, P., Shamovsky, O., Palma, A, Mikkelsen, T.S., Altshuler, D., and Lander, E.S. (2006). Positive natural selection in the human lineage. Science *312*, 1614–1620.

Samson, M., Libert, F., Doranz, B.J., Rucker, J., Liesnard, C., Farber, C.M., Saragosti, S., Lapoumeroulie, C., Cognaux, J., Forceille, C., *et al.* (1996). Resistance to HIV-1 infection in caucasian individuals bearing mutant alleles of the CCR-5 chemokine receptor gene. Nature *382*, 722–725.

Sandaa, R.-A., Gómez-Consarnau, L., Pinhassi, J., Riemann, L., Malits, A., Weinbauer, M.G., Gasol, J.M., and Thingstad, T.F. (2009). Viral control of bacterial biodiversity – evidence from a nutrient-enriched marine mesocosm experiment. Environ. Microbiol. *11*, 2585–2597.

Sanz, Y., Santacruz, A., and De Palma, G. (2008). Insights into the roles of gut microbes in obesity. Interdiscip. Perspect. Infect. Dis. *2008*, 829101.

Savage, D.C. (1977). Microbial ecology of the gastrointestinal tract. Ann. Rev. Microbiol. *31*, 107–133.

Sawyer, S.L., Emerman, M., and Malik, H.S. (2004). Ancient adaptive evolution of the primate antiviral DNA-editing enzyme APOBEC3G. PLoS Biol. *2*, E275.

Sawyer, S.L., Wu, L.I., Emerman, M., and Malik, H.S. (2005). Positive selection of primate TRIM5alpha identifies a critical species-specific retroviral restriction domain. Proc. Natl. Acad. Sci. U.S.A. *102*, 2832–2837.

Schulte, R.D., Makus, C., Hasert, B., Michiels, N.K., and Schulenburg, H. (2010). Multiple reciprocal adaptations and rapid genetic change upon experimental coevolution of an animal host and its microbial parasite. Proc. Natl. Acad. Sci. U.S.A. *107*, 7359–7364.

Schwiertz, A., Gruhl, B., Löbnitz, M., Michel, P., Radke, M., and Blaut, M. (2003). Development of the intestinal bacterial composition in hospitalized preterm infants in comparison with breast-fed, full-term infants. Pediatr. Res. *54*, 393–399.

Shafer, W.M., and Ohneck, E.A. (2011). Taking the gonococcus–human relationship to a whole new level: implications for the coevolution of microbes and humans. mBio *2*, e00067–11.

Shapiro, O.H., Kushmaro, A., and Brenner, A. (2010). Bacteriophage predation regulates microbial abundance and diversity in a full-scale bioreactor treating industrial wastewater. ISME J. *4*, 327–336.

Shih, A.C.-C., Hsiao, T.-C., Ho, M.-S., and Li, W.-H. (2007). Simultaneous amino acid substitutions at antigenic sites drive influenza A hemagglutinin evolution. Proc. Natl. Acad. Sci. U.S.A. *104*, 6283–6288.

Shoemaker, N.B., Vlamakis, H., Hayes, K., and Salyers, A.A. (2001). Evidence for extensive resistance gene transfer among Bacteroides spp. and among Bacteroides and other genera in the human colon. Appl. Environ. Microbiol. *67*, 561–568.

Smillie, C.S., Smith, M.B., Friedman, J., Cordero, O.X., David, L.A., and Alm, E.J. (2011). Ecology drives a global network of gene exchange connecting the human microbiome. Nature *480(7376)*, 241–244.

Stanton, M.L. (2003). Interacting guilds: moving beyond the pairwise perspective on mutualisms. Am. Nat. *162*, S10–S23.

Stark, P.L., and Lee, A. (1982). The microbial ecology of the large bowel of breast-fed and formula-fed infants during the first year of life. J. Med. Microbiol. *15*, 189–203.

Stephens, J.C., Reich, D.E., Goldstein, D.B., Shin, H.D., Smith, M.W., Carrington, M., Winkler, C., Huttley, G.A., Allikmets, R., Schriml, L., *et al.* (1998). Dating the origin of the CCR5-Delta32 AIDS-resistance allele by the coalescence of haplotypes. Am. J. Hum. Genet. *62*, 1507–1515.

Thompson, J. (1994). The Coevolutionary Process (University of Chicago Press, Chicago, IL).

Thompson, J. (2005a). Coevolution: the geographic mosaic of coevolutionary arms races. Curr. Biol. *15*, R992–R994.

Thompson, J. (2005b). The Geographic Mosaic of Coevolution (University of Chicago Press, Chicago, IL).

Vallender, E.J., and Lahn, B.T. (2004). Positive selection on the human genome. Hum. Mol. Genet. *13*, R245–R254.

Van Valen, L. (1973). A new evolutionary law. Evol. Theory *1*, 1–30.

Verberkmoes, N.C., Russell, A.L., Shah, M., Godzik, A., Rosenquist, M., Halfvarson, J., Lefsrud, M.G., Apajalahti, J., Tysk, C., Hettich, R.L., *et al.* (2009). Shotgun metaproteomics of the human distal gut microbiota. ISME J. 3, 179–189.

Waters, V.L. (2001). Conjugation between bacterial and mammalian cells. Nat. Gen. *29*, 375–376.

Whitman, W.B., Coleman, D.C., and Wiebe, W.J. (1998). Prokaryotes : The unseen majority. Proc. Natl. Acad. Sci. USA. *95*, 6578–6583.

Wilmes, P., Remis, J.P., Hwang, M., Auer, M., Thelen, M.P., and Banfield, J.F. (2009). Natural acidophilic biofilm communities reflect distinct organismal and functional organization. ISME J. *3*, 266–270.

Wilson, D., and Sober, E. (1989). Reviving the superorganism. J. Theor. Biol. *136*, 337–356.

Xu, J., Mahowald, M.A., Ley, R.E., Lozupone, C.A., Hamady, M., Martens, E.C., Henrissat, B., Coutinho, P.M., Minx, P., Latreille, P., *et al.* (2007). Evolution of symbiotic bacteria in the distal human intestine. PLoS Biol. *5*, e156.

Yang, X., Xie, L., Li, Y., and Wei, C. (2009). More than 9,000,000 unique genes in human gut bacterial community: estimating gene numbers inside a human body. PloS ONE *4*, e6074.

Zaneveld, J.R., Lozupone, C., Gordon, J.I., and Knight, R. (2010). Ribosomal RNA diversity predicts genome diversity in gut bacteria and their relatives. Nucleic Acids Res. *38*, 3869–3879.

Zell, R. (2004). Global climate change and the emergence/re-emergence of infectious diseases. Int. J. Med. Microbiol. *293*, 16–26.

Zhang, J., and Webb, D.M. (2004). Rapid evolution of primate antiviral enzyme APOBEC3G. Hum Mol Genet. *13*, 1785–1791.

Zheng, X., Xie, G., Zhao, A., Zhao, L., Yao, C., Chiu, N.H.L., Zhou, Z., Bao, Y., Jia, W., Nicholson, J.K., *et al.* (2011). The footprints of gut microbial-mammalian co-metabolism. J. Proteome Res. 10, 5512–22.

Zilber-Rosenberg, I., and Rosenberg, E. (2008). Role of micro-organisms in the evolution of animals and plants: the hologenome theory of evolution. FEMS Microbiol. Rev. *32*, 723–735.

Zoetendal, E.G., Akkermans, A.D., and De Vos, W.M. (1998). Temperature gradient gel electrophoresis analysis of 16S rRNA from human fecal samples reveals stable and host-specific communities of active bacteria. Appl. Environ. Microbiol. *64*, 3854–3859

Mutualism: Plant–Microorganism Interactions

3

Penny R. Hirsch and Tim H. Mauchline

Abstract

Mutualism is responsible for the genesis of green plants and is implicated in their colonization of land. Current knowledge of plant–microorganism symbioses includes a range of associations with different degrees of intimacy and mutual dependence but the mutual benefits are not always clear. Complex signalling is involved when the plant immune system recognizes beneficial endosymbionts although many have also evolved mechanisms to evade or moderate plant defence pathways. A wide range of bacteria inhabit intercellular spaces but only a few are true endosymbionts able to penetrate living cells whilst remaining membrane-bound, accessing plant carbon compounds in a manner analogous to biotrophic pathogens. Unlike pathogens, they provide nutrients to the plant in exchange. The best-known examples are rhizobia, bacteria that induce root nodules on leguminous plants and fix atmospheric nitrogen; and arbuscular mycorrhizal fungi that sequester phosphate and organic N from soil and provide it to their plant hosts. Both secrete factors prior to contacting plant cells which appear to prepare the hosts for mutual rather than pathogenic interactions and suppress the defence mechanisms. The processes involved in these symbioses are compared with less intimate interactions and the nature of mutualism is discussed.

Introduction

Microorganisms have evolved to inhabit myriad niches with a range of lifestyles, from chemoautotrophy to biotrophy. Plants, relative newcomers to terrestrial life, do not live in isolation, but rather as primary producers, they are central to ecosystem functioning. Above ground interactions, be it with animals or other plants, are relatively well understood in comparison with those below ground. Root systems are known to interact in many different associations with the soil microbiota; it is likely that many others remain to be discovered when we consider that there are 10^9 soil microbes and at least 10,000 species in 1 g soil (Gans *et al.*, 2005). Bacteria and fungi are the most numerous microorganisms present in soil. The mobility of bacteria in soil is limited, and although chemotaxis to root exudates may be important on a micro-scale, most interactions are initiated when plant roots growing through soil come into contact with bacteria adhering to aggregates and in water films. Fungal hyphae, however, can traverse the matrix of mineral and organic matter aggregates and voids that constitute the soil matrix.

If they can utilize substrates exuded or deposited by roots, saprophytic bacteria and fungi will proliferate in the rhizosphere and colonize the root surface, as shown in Fig. 3.1a and d

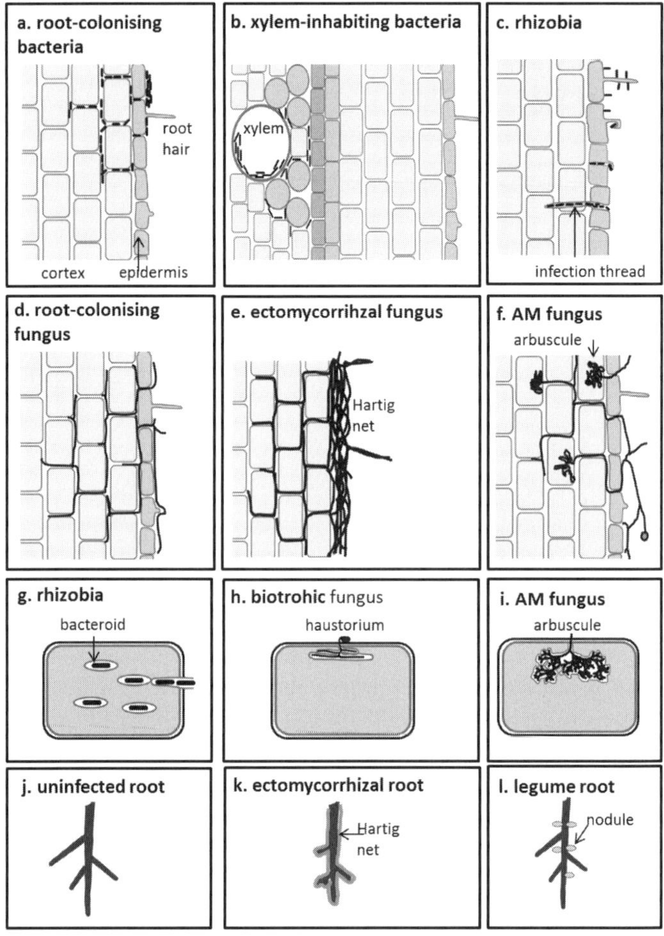

Figure 3.1 Plant–microorganism associations. The plant cells, bacteria (black rods) and fungal hyphae (black lines) are not drawn to scale. (a) Many soil bacteria can colonize the root surface and outer cortical cells. (b) It is more difficult for bacteria to reach the inner root. Bacteria that inhabit xylem are often fastidious. Some pathogens are found in phloem (depicted as round cells) and xylem, often transmitted by sap-sucking insects. (c) Stimulated by plant signals, rhizobia secrete Nod factors and attach to root hairs initiating curling and infection thread formation. This penetrates the cortex and is colonized by rhizobia. (d) Many soil fungi colonize root surfaces and grow between cortical cells. (e) Ectomycorrhizal fungi form and extensive hyphal network around roots, the Hartig net. (f) Prior to root colonization, Myc factors secreted by arbuscular mycorrhizal fungi in response to plant signals prepare the cortical cells for penetration by fungal hyphae and arbuscule formation. (g) As the root nodule forms, rhizobia are budded off from the infection thread in a 'peribacteroid membrane' of plant origin which surrounds the bacterial cell as it differentiates into the nitrogen-fixing bacteroid. (h) Biotrophic fungal pathogens penetrate living cells and form haustoria, feeding structures that are bounded by the plant cell membrane. (i) The AM fungal arbuscules are sometimes less distinctive and resemble haustoria but are always surrounded by the plant cell membrane. (j) Most microorganisms that interact with roots do not cause morphological changes although production of phytohormones can increase root mass. (k) Ectomycorrhizal roots are sheathed by the Hartig net which protects them and provides nutrients. It also causes morphological changes, typically shorter roots with terminal bifurcations. (l) The root nodules induced on legume roots by rhizobia each contain 10^6–10^9 viable bacterial cells in addition to the bacteroids which lose the ability to divide.

(Dennis *et al.*, 2010). Only a minority of soil microorganisms that come into contact with roots or the above-ground plant will have more specific relationships, either as pathogens causing damage or as symbionts living in association with the host. These may be obligate biotrophs that grow only on living cells, or be capable of independent saprophytic growth. Symbiotic lifestyles range from parasitism (one partner benefits to the detriment of the other); commensalism (one partner benefits and the other is not affected); or mutualism (both partners benefit). However, in nature such discretely defined boundaries rarely apply: parasitism and mutualism represent ends of an uninterrupted continuum in which the status of symbiotic associations can shift (Douglas, 2008).

There are well-known mutualistic relationships in which soil microorganisms enhance the nutrient supply to plants, which will be discussed in detail below. One example is the relatively specific association between the nitrogen-fixing bacteria commonly known as rhizobia and legumes – plants belonging to the family Fabaceae. Another involves the Glomeromycota, fungi that form arbuscular mycorrhizas with many plant species and improve uptake of phosphate and other nutrients. In the established examples of N-fixing symbioses, mutualism has been verified by demonstrating that atmospheric N_2 is incorporated into the plant host and that the bacterial endosymbionts are essential for this to occur. Many other symbioses between plants and endophytic bacteria that are potentially capable of fixing N_2 have been described but there is rarely good evidence that these are truly mutualistic: the contribution of biological nitrogen fixation (BNF) to the plant's nutritional status is often questionable and the benefits difficult to verify (Herridge *et al.*, 2008). Nevertheless there is considerable interest in such associations with crop plants as they offer the possibility of reducing nitrogen fertilizer inputs to agriculture.

However, there are likely to be many mutualistic associations that have not yet been recognized, for example where the microorganisms cannot be readily identified or grown in laboratory culture. The recent developments of high-throughput 'next generation' sequencing technologies and associated 'catch-all' metagenomic tools will provide the opportunity to detect the presence and function of as yet cryptic microorganisms in symbiotic association with unprecedented resolution. They will also greatly improve our understanding of the biodiversity and ecology of partners in the more familiar mutualistic associations. With an expanding global population and the consequent pressures on agricultural production and constraints on fertilizer availability, mutualistic associations between crop plants and soil microorganisms are likely to become increasingly important.

Who are the partners?

Overview

Plants have mutualistic relationships with some animals, especially insects (more rarely, vertebrates) usually involving pollination or seed dispersal (Bronstein *et al.*, 2006; Fleming and Muchhala, 2008). Protists, together with a range of other microorganisms, play a mutualistic role by enhancing degradation of trapped prey in pitcher plants (Adlassnig *et al.*, 2011). Obligate endoparasitic protists include oomycetes (previously classified as fungi), the causative agents of some serious plant diseases including *Phytophthora infestans* which is responsible for potato blight, *Pythium* spp. that cause damping off, and also some rusts and downy mildews. The plasmodiophorids are another group of endoparasitic protists that cause club root in brassicas and act as vectors for viral diseases in other plants. They can form asymptomatic associations (Smith *et al.*, 2011) but no other examples of protist–plant mutualism have been described.

Ammonia oxidizers belonging to both prokaryotic domains, the Bacteria and the Archaea, are ubiquitous in soil. They play an important role in nitrification, the process by which relatively insoluble and immobile ammonia is converted to nitrate which is more available to plants (and to leaching losses). Although ammonia-oxidizing archaea appear predominant in many soils, their relative contribution to nitrification is uncertain. Archaea have been detected in rhizosphere communities (Mendes *et al.*, 2011) but there is no evidence to date that they form mutualistic associations with plants. This contrasts to the large body of research on mutualistic bacteria and fungi.

Bacteria

BNF, mediated by the nitrogenase enzyme, is found only in prokaryotes. Nitrogen is an essential component of life and nitrogen gas is abundant but plants cannot access this atmospheric nitrogen directly, and in the absence of man-made fertilizers, all terrestrial life is ultimately dependent on BNF. However this is an energy-intensive process and is limited by the availability of carbon substrates in soil. Conversely, green plants have access to solar energy and generate carbon compounds via photosynthesis, providing the basis for the many mutualistic associations between nitrogen-fixing bacteria and plants. Bacteria forming nitrogen-fixing mutualisms include members of the phyla Proteobacteria, Actinobacteria and Cyanobacteria. The rhizobia, the best-known examples, induce and inhabit root nodules on a range of plants including agronomically important crops such as clovers, soybeans, lentils, beans and peas. They are also important in natural ecosystems, in particular nutrient-poor soils. The interactions are highly specific, with only the 'correct' rhizobia forming nitrogen-fixing symbioses with a particular host plant. Most belong to the order Rhizobiales in the α-Proteobacteria (α-rhizobia) although recently β-Proteobacteria from the order Burkholderiales have been identified, now referred to as β-rhizobia (Masson-Boivin *et al.*, 2009).

Frankia are a separate group of nitrogen-fixing root-nodule bacteria. They belong to the Actinobacteria and associate with shrubs and trees, many of which are pioneer species that grow on marginal land. Actinorhizal plants belong to eight different families in four different orders (Benson *et al.*, 2011). This contrasts to the rhizobia–legume symbiosis where plant hosts are restricted to the Fabaceae, with the notable exception of *Parasponia*, a genus of the family Ulmaceae (Gough and Cullimore, 2011).

Cyanobacteria–plant associations have evolved in a range of vascular plants with a variety of specialized structures. The aquatic fern *Azolla* associates with *Anabaena* spp.; cycads form 'coralloid' roots containing several genera, predominantly *Nostoc*, which also associates with the flowering plant *Gunnera* (Vessey *et al.*, 2004). Although *Gunnera* is primarily known as ornamental 'giant rhubarb', *Azolla* is grown in paddies to fertilize rice and cycads are cultivated as crop plants in some cultures, so these cyanobacterial mutualisms are of agronomic as well as ecological importance.

There are many examples of symbiotic associations between nitrogen-fixing bacteria and plants where mutualism has not been verified. Some of these are difficult to work with, hard to isolate and culture; with resemblance to certain vascular pathogens in particular those inhabiting xylem as indicated in Fig. 3.1b (Bove and Garnier, 2002). A few are better known: the bacterial genome has been sequenced in an *Azoarcus* sp. that inhabits Kallar grass (Hurek *et al.*, 2002; Krause *et al.*, 2006), *Herbaspirillum seropedicae* associated with a range of tropical grasses (Gyaneshwar *et al.*, 2002; Pedrosa *et al.*, 2011) and *Gluconacetobacter diazotrophicus* associated with sugar cane (Sevilla *et al.*, 2001; Bertalan *et al.*, 2009). However, mutants of these bacteria incapable of BNF also have beneficial effects on plant growth and analysis of the bacterial genomes indicates that they also synthesize growth-promoting phytohormones and may enhance resistance to invasion by pathogens (Rosenblueth and Martinez-Romero, 2006). This may also be the case for a range of bacteria including Proteobacteria, Firmicutes and Actinomycetes that colonize the root surface and the outer root cortex as shown in Fig. 3.1a (Rosenblueth and Martinez-Romero, 2006), where some are capable of BNF. Although often assumed to be advantageous to the plant hosts, there is scant evidence that BNF in these exterior locations benefits the plant, rather it confers a competitive advantage on the bacterial colonists and it is debatable if such interactions should be described as mutualistic. However, there is also considerable interest in the plant growth-promoting potential of these bacteria (Compant *et al.*, 2010). Well-studied examples include the α-proteobacterial genus *Azospirillum* (Franche *et al.*, 2009) and the γ-proteobacterium *Azotobacter paspali* (Lugtenberg and Kamilova, 2009), both capable of BNF. However they also include non-N-fixing bacteria, in particular *Pseudomonas fluorescens* (γ-Proteobacteria) and *Bacillus* spp. (Firmicutes), which (in common with many other groups) are also implicated in improving host plant resistance to pathogens (Lugtenberg and Kamilova, 2009).

Fungi

Mycorrhiza means 'fungus-root' and mycorrhizal fungi form mutualistic symbiotic associations with roots, their hyphae extending into soil and effectively increasing the root surface available for nutrient and water uptake and in return obtaining sugars (Brundrett, 2002; Finlay, 2008). There are three main groups: ectomycorrhizal fungi from several phyla (Basidiomycota, Ascomycota, Zygomycota) sheath the roots of trees and shrubs (Fig. 3.1e and k); ericoid mycorrhizal fungi are Ascomycota associated with ericaceous plants; and endomycorrhizal fungi that form symbioses with a wide range of plant groups. In contrast to the other two groups, endomycorrhizal fungi are obligate symbionts belonging to the Glomeromycota and are commonly referred to as arbuscular mycorhizal fungi (AMF) on account of the arbuscules (branching hyphal structures) formed within host root cortical cells (Fig. 3.1f and i). The majority of angiosperm phyla (> 80%) are reported to form AMF, which are seen in fossils and are thought to have co-evolved with land plants, whereas non-mycorrhizal plant families (including the Chenopodiaceae and Brassicaceae) appear to actively exclude fungal infection (Brundrett, 2002). In natural ecosystems, the majority of plants are found to be mycorrhizal but agricultural cultivation reduces infection levels, in part because tillage breaks up mycelial networks in soil, and plants replete in N and P may be less susceptible to AMF colonization although there is considerable variation between species (Treseder and Allen, 2002). Roots colonized by AMF take up less phosphate directly from soil than uninfected plants, instead it is taken up by the hyphae and transported to the plant, indicating that AMF may suppress root alkaline phosphatase and phosphate transporter activity (Smith *et al.*, 2003). Fungal alkaline phosphatase is active in internal hyphae in roots (Tani *et al.*, 2009), much less so in extraradical hyphae (Aono *et al.*, 2004). The consequences of these changes in root physiology arising from AMF symbiosis is an overall improvement in mineral acquisition, most notably P, by the plant host.

Mycorrhizal associations are generally assumed to be mutualistic but the symbioses between orchids and Basidiomycota is complex. All orchid seeds are minute, containing insufficient carbon to nourish the germinating plant and at this stage a fungal partner is essential, providing organic carbon from soil organic matter. Some non-photosynthetic orchids rely on saprophytic fungi for nutrition throughout their life, and some ectomycorrhizal fungi associate with both trees and orchids simultaneously, providing an indirect link between unrelated plant species. However, in some cases the fungal partner becomes parasitic rather than mutualistic, reducing rather than supporting growth of the orchid (Rasmussen, 2002). This illustrates that 'mutualistic' relationships are ultimately selfish, with each partner ultimately protecting their own survival.

Where do microorganisms reside?

Nitrogen-fixing bacteria

For effective BNF, microorganisms require abundant energy and carbon. Many are obligate aerobes although nitrogenase is irreversibly inactivated by oxygen and in rhizobia, BNF is suppressed at atmospheric concentrations. The root nodules induced and inhabited by rhizobia or frankia and the coralloid roots occupied by cyanobacteria offer a suitable haven with preferential access to carbon substrates for the bacteria, excluding competitors present in the rhizosphere, providing nitrogen compounds to the plant and limiting oxygen supply (Vessey *et al.*, 2004). The rhizobia–legume symbiosis is the best studied, with a cascade of signalling events identified. In many cases, rhizobia enter an invagination they induce in root hairs, that develops into an infection thread (Fig. 3.1c). This grows into the root cortex and as rhizobia proliferate in the thread, a meristem is induced to produce the nodule into which rhizobia are released, enclosed in plant plasma membrane-bound vesicles, where they mature into nitrogen-fixing bacteroids shown in Fig. 3.1g (Franche *et al.*, 2009). Another mode of infection is through cracks where lateral roots emerge, with subsequent induction of a nodule meristem and formation of bacteroids (Masson-Boivin *et al.*, 2009). Frankia and cyanobacteria form thick-walled vesicles or heterocysts in roots where BNF occurs, further limiting the exposure of nitrogenase to O_2. In legume nodules, rhizobia differentiate into large N-fixing bacteroids and O_2 is provided by the carrier protein leghaemoglobin comprising globin synthesized by the legume host and a haem moiety from the bacterial partner (Franche *et al.*, 2009). A similar strategy is present in foliar associations between cyanobacteria and plants: *Anabaena* occupies leaf cavities in *Azolla*; *Nostoc* inhabits glands at the petiole bases in *Gunnera*, N-fixation is confined to the heterocysts that develop (Franche *et al.*, 2009).

These truly mutualistic associations involve specialized plant structures that house the bacterial partners and support efficient rates of BNF. In agricultural systems, rhizobia in legumes are estimated to fix annually 110–225 kg N per ha, contrasting with 25 kg N per ha for sugar cane and < 10 kg N per ha in other endophytic and rhizosphere associations (Herridge *et al.*, 2008). The endophytic bacteria colonize intracellular spaces and vascular tissues (Fig. 3.1b) above and below ground: the lower rates of BNF may indicate that the relative number of bacteria to plant tissue is less in these associations, conditions are less conducive for nitrogenase activity and/or that fixed N is released to the plant only after the death of the endophyte.

Mycorrhizal fungi

Mycorrhizal fungi colonize the root surface and hyphae grow into the root, between cortical cells. As obligate symbionts AMF survive as large spores but cannot proliferate until a compatible host is encountered. Hyphae emerge sporadically from the spore and withdraw if there are no suitable roots in proximity. They penetrate cortical cells whilst remaining surrounded by a plant plasma membrane, developing the typical arbuscules (Fig. 3.1f and i) which are the sites of nutrient exchange: the AMF are particularly associated with improving phosphate supply to the host plant, but also enhance provision of other nutrients and water (Parniske, 2008). The fungal endosymbionts of orchids are capable of independent saprophytic existence in soil, but germinating seeds are dependent on the fungi, which invade roots and penetrate cortical cells where nutrient transfer may involve the breakdown of fungal tissue (Rasmussen, 2002).

Ericoid and ectomycorrhizal fungi are also capable of independent growth in soil and are important in transferring both organic nitrogen and phosphate from soil to plant host. Ericoid endomycorrhizal fungi penetrate root cells but do not develop arbuscules, whereas the ectomycorrhizal fungi colonize intercellular spaces without penetrating root cortical cells forming a structure called the Hartig net shown in Fig. 3.1e and k (Brundrett, 2004). In contrast, the membrane-bound hyphal structures that penetrate root cells in endomycorrhizas resemble the haustoria of biotrophic fungal pathogens (i.e. those feeding on living plant cells) as shown in Fig. 3.1h.

Other plant-associated microorganisms

There is great competition between the multifarious microorganisms in soil, which is nutrient-poor, to colonize roots and other plant surfaces that exude carbon substrates. Many such interactions are generally advantageous to the plant as they limit pathogenic microorganisms either by competitive exclusion or production of antagonistic compounds, some confer specific benefits such as synthesis of phytohormones that stimulate root growth (Lugtenberg and Kamilova, 2009). Bacteria and fungi in these interactions tend to proliferate in regions where there is active root exudation such as around a germinating seed, or where older cells lyse in the basal region (Dennis *et al.*, 2009). Above-ground plant surfaces, known collectively as the phyllosphere, are also colonized by microorganisms, although it is a more changeable environment than soil, with diurnal light and temperature fluctuations and rapid changes in temperature and moisture. The density of microorganisms in the phyllosphere is proportionally less than below ground, reflecting this harsher environment. Nevertheless, some phyllosphere microorganisms are beneficial to the plant, providing protection against pests, pathogens or environmental changes (Lindow and Brandl, 2003).

Endophytes have been found in all species of plant studied to date in a wide range of niches within the host, such as roots, stems, leaves, seeds, fruits, tubers, ovules, although bacterial numbers never approach those of rhizobia found in legume nodules. Similarly, endophytic fungal growth is never as profuse as that of ecto- or endomycorrhizal fungi. Generally more are found in below- rather than above-ground parts of the host but unlike rhizobia, they are not found as membrane-bound intracellular inclusions (Rosenblueth and Martinez-Romero, 2006). Beneficial effects have been ascribed to some but their ubiquity makes this difficult to assess in many cases. It is necessary to demonstrate a plant is free of endophytes, and then reintroduce specific cultures. However, endophytes can be seed-borne, and some crops such as sugar cane are subject to vegetative propagation, so it can be extremely difficult to generate endophyte-free plant material, and many of the bacteria cannot be isolated in pure culture.

Fungal endophytes that colonize root intercellular spaces may protect against fungal pathogens or nematode pests, by parasitism and/or production of antagonistic compounds. In many grasses, seed-borne endophytic fungi develop systemically and may produce toxins that protect against herbivory, e.g. *Epichloë festucae* in perennial ryegrass. However, the fungi may also reduce seed production (Schardl *et al.*, 2004) – at certain growth stages of the host, *Epichloë* switches to effectively becoming a plant pathogen (Eaton *et al.*, 2011). Although these examples are based on mutual exploitation, there are many similarities between the fungal endophytes and plant pathogens (Saikkonen *et al.*, 2004).

How do microorganisms colonize plant tissues and evade defences?

Pathogens and other endophytes

In the absence of the adaptive, circulating immune systems of animals, plants rely on innate immunity as a first line of defence (Spoel and Dong, 2012). During pathogenesis microbes enter plant tissues through natural openings, wounds, enzymatic disruption and in the case of some fungi, physical pressure imposed by appressoria which develop on external plant surfaces and enable hyphae to penetrate cells or intercellular spaces, by a number of different infection strategies: Nectrotrophic pathogens invade dead or damaged tissue, biotrophs feed on living cells and hemibiotrophs ultimately kill cells (Hammond-Kosack and Jones, 2000). Plants can recognize different invading microorganisms and cellular damage and respond by increasing local defences, localized cell death to limit the spread of infection, and induction of systemic resistance involving volatile signals (Uma *et al.*, 2011). This 'non-host resistance' is initiated when molecules associated with microorganisms are recognized by plant receptors: these may be generic signals or typical microbial- or pathogen-associated molecular patterns known as MAMPs and PAMPs (Jones and Dangl, 2006). The pattern-triggered immunity (PTI) involves release of reactive oxygen that may kill the invader, generate localized cell-wall thickening, and lead to induced systemic resistance (ISR) mediated by transient synthesis of ethylene and jasmonic acid. The PTI can be overcome by pathogens that deploy virulence factors or 'effectors' which may in turn be recognized by the plant resistance (R) proteins. This leads to a more robust effector-triggered immunity (ETI) that results in localized cell death (a hypersensitive response or HR) and localized and systemic acquired resistance (SAR) mediated by salicylic acid that protects against further infections (Spoel and Dong, 2012). The pathogen virulence genes can evolve rapidly to generate progeny with effectors that evade plant defences, whereas plants evolve resistance to overcome effectors (Hammond-Kosack and Jones, 2000). This differs from the MAMPs and PAMPs that represent core features of microorganisms such as fungal chitin, and elements that evolve slowly, such as flagellin (Jones and Dangl, 2006). The endless evolutionary arms race of pathogen virulence and plant resistance contrasts with the mutually beneficial associations seen in the rhizobia–legume and mycorrhizal symbioses which have not been observed to change rapidly over time.

Sequencing the genome of the successful endosymbiont *Azoarcus* revealed that it lacks the secretion systems through which related pathogenic bacteria deliver virulence factors to the plant cell (Krause *et al.*, 2006). These would normally be identified as PAMPs, but the extent to which endophytes are 'disarmed pathogens' is unclear, for example flagella may be required for efficient plant colonization, but will also be recognized as PAMPs, and plants may show limited defence responses to non-pathogenic endophytes (Reinhold-Hurek and Hurek, 2011).

Mutualistic endophytes

Host plant immune responses can potentially damage endophytes, especially those in endosymbiotic associations. Exposure of plant cell membranes to MAMPs originating from microorganisms that colonize plant surfaces or inhabit intercellular spaces is less than those interacting at the cell surface. Nevertheless, roots react strongly to bacterial and fungal MAMPs (Millet *et al.*, 2010). Successful endophytes may lack some MAMPs or PAMPs and can elude recognition by relatively rapid changes, e.g. the phenotypic variation observed in endophytic pseudomonads (Zamioudis and Pieterse, 2011). Phase variation is well known in animal pathogens, where bacterial replication induces reversible genetic or epigenetic changes in cell surface features such as flagella and lipopolysaccharides (LPS). Nevertheless, rhizosphere microorganisms and endophytes that elicit PTI and generate ISR without obvious deleterious effects are seen as beneficial mutualists (Lugtenberg and Kamilova, 2009). Some also produce effector molecules to suppress salicylic acid signalling and ETI, and these may be involved in conferring a degree of specificity (Zamioudis and Pieterse, 2011).

These strategies are also seen in endophytes that form more intimate associations with cells, but they have evolved additional means of overcoming PTI and ETI. The association between plants and AMF is thought to predate other mutualistic symbioses, which appear to require some of the same plant genes and to involve similar pathways, albeit with differences in the signalling (Parniske, 2008). Plants release strigolactones, which are elevated in potential hosts especially when phosphate-starved (Paszkowski, 2006). These stimulate growth and branching of fungal hyphae and release of the fungal 'Myc' factor, a small diffusible molecule unidentified to date (Bucher *et al.*, 2009). Less is known about further molecular signalling between host and AMF but the processes involved are well described. On encountering the root surface, hyphae proliferate by increased branching and colonize the intercellular spaces of the inner cortex where they penetrate cells to form the membrane-bound arbuscules. The plant genes activated during this process are also up-regulated during infection by biotrophic fungi, indicating that AMF effectors suppress ETI; similarly, the Myc factors are thought to suppress PTI. Ectomycorrhizal fungi are also perceived initially by their plant host as pathogenic invaders, elicit a PTI that is subsequently suppressed by effectors (Zamioudis and Pieterse, 2011).

Rhizobia that form root nodules via an infection thread secrete lipochitooligosaccharide Nod-factors in response to flavonoids exuded by their host plant. The Nod factors trigger the steps that lead to nodule formation and the generation of nitrogen-fixing bacteroids (Masson-Boivin *et al.*, 2009). This is a complex, highly evolved interaction that requires specific recognition by both bacterial and plant partner, yet rhizobia have many features such as flagella and secretion systems that function as MAMPs or PAMPs. However, it appears that rhizobial lipo- and exopolysaccharides, which can induce PTI in non-host plants, actively suppress it in host legumes. Furthermore, both 'incorrect' rhizobia and mutants of normally compatible strains deficient in exopolysaccharides induce PTI and increase salicylic acid levels in the plant (Zamioudis and Pieterse, 2011) whereas the LPS of the compatible alfalfa symbiont *Sinorhizobium meliloti* appears to suppress PTI and ETI (Tellstroem *et al.*, 2007). The suppression induced by such compatible rhizobia appears to be generated by effectors analogous to those produced by pathogens, and is thought to contribute to the specificity of the interaction, the establishment of the infection, development of nodules and ultimately the success of the mutualism (Zamioudis and Pieterse, 2011).

Both Nod and Myc factors induce a calcium oscillation in the cells of compatible host plants which appears to prepare the cells for establishment of the symbiosis and is different from that induced by pathogens (Oldroyd *et al.*, 2009).

The origins of mutualism

Green plants are believed to contain the relics of two mutualisms that occurred during early evolution. Mitochondria, the energy-generating organelles of all eukaryotes, are thought to be derived from an endosymbiotic α-Proteobacterium whereas chloroplasts arose from a subsequent endosymbiosis with cyanobacteria (Keeling, 2010). Both of these have greatly reduced genomes, relying on the plant nucleus for their continued maintenance and function. The chloroplasts provide carbon and energy to plants, which are estimated to lose up to 40% of photosynthetically derived carbon via their roots, with exudation and rhizodeposition supporting the rhizosphere community (Lynch and Whipps, 1990). Mutualistic endophytes may provide both partners with substantial benefits whilst utilizing some of this substrate. It is estimated that up to 20% of photosynthetically derived carbon is diverted to nodules containing rhizobia (Minchin and Pate, 1973) and a similar amount is apportioned to AMF and other mycorrhizal fungi (Bago *et al.*, 2002). Of these, only AMF are totally dependent on the endomycorrhizal association for proliferation and survival. The longevity of this association, believed to have evolved more than 400 million years ago when land plants evolved, and its ubiquity (70–90% of all land plant species) provide evidence of its importance to the hosts (Parniske, 2008). Ectomycorrhizal fungi, whilst not penetrating plant cells, modify the roots of the host plant significantly (Fig. 3.1j and k). Whilst legumes acquire a dedicated source of nitrogen, the rhizobia modify root structure by initiating root nodule development and gain a protected niche in which they can proliferate (Fig. 3.1g, j and l). Although cells that differentiate into nitrogen-fixing bacteroids lose the ability to divide, they comprise only a minority of the total nodule community, with at least 10^6 and up to 10^9 rhizobial cells (depending on the plant species and nodule size) released into soil when each nodule senesces and an elevation in the numbers in soil of several orders of magnitude following a legume crop (Hirsch, 1996). Therefore, even in a mutualism with demonstrable benefits to both partners, a large proportion of the microbial symbionts are 'free riders' that do not directly benefit the host (Denison *et al.*, 2003). However these are essential for the survival of the microbial partners in soil, to nodulate the next generation of plant hosts, thus represent an essential step in the mutualistic association.

Perspective

Our experience of working with both mycorrhizal and rhizobia-nodulated plants in artificial light illustrates the potential drain of the endophyte on the host. If light levels drop below a critical threshold, plants rapidly become sick and resemble those suffering a pathogenic infection. Some rhizobia perform suboptimally in certain hosts and mutants lacking the genes for nitrogenase may still induce and inhabit nodules. However plants exercise some control over suboptimal rhizobial endophytes and reduce resources provided to the nodule (Denison and Kiers, 2011). Also, the ability of rhizobia to form nodules is greatly reduced in soils replete with plant-available nitrogen, mediated by a systemic regulator. Although the precise nature of this is not yet clear, plant mutants in genes regulating nodule number become 'super-nodulating' and roots resemble those suffering a serious pathogen infection (Reid *et al.*, 2011). Similarly, AMF infection is inhibited when levels of nitrogen and phosphate are high. Thus despite the potential for free-riders or cheats discussed previously, both partners benefit in the long-established and well-studied rhizobia and AMF–plant associations and they are truly mutualistic. It is less clear in some other cases: dramatic switches from beneficial symbiont to pathogen have been described in basidiomycete–orchid and *Epichloë*–grass associations (Rasmussen, 2002; Eaton *et al.*, 2011). This review illustrates the continuum of plant–microbe associations, from those involving highly specificity and the modification of the plant to provide a niche, to more ephemeral partnerships is which the description of 'mutual' may be more subjective. Increased knowledge concerning the nature of signalling pathways and the consequences of mutations will illuminate this fascinating topic. It may then become possible to modify endophytes to improve the nutrient acquisition, disease resistance and growth characteristics of economically important plants leading to truly sustainable agriculture.

Acknowledgement

Rothamsted Research receives strategic funding from the Biotechnology and Biological Sciences Research Council.

References

Adlassnig, W., Peroutka, M., and Lendi, T. (2011). Traps of carnivorous pitcher plants as a habitat: composition of the fluid, biodiversity and mutualistic activities. Ann. Bot. *107*, 181–194.

Aono, T., Maldonado-Mendoza, I.E., Dewbre, G.R., Harrison, M.J., and Saito, M. (2004). Expression of alkaline phosphatase genes in arbuscular mycorrhizas. New Phytol. *162*, 525–534.

Bago, B., Zipfel, W., Williams, R.M., Jun, J., Arreola, R., Lammers, P.J., Pfeffer, P.E., and Shachar-Hill, Y. (2002). Translocation and utilization of fungal storage lipid in the arbuscular mycorrhizal symbiosis. Plant Physiol. *128*, 108–124.

Benson, D.R., Brooks, J.M., Huang, Y., Bickhart, D.M., and Mastronunzio, J.E. (2011). The biology of *Frankia* sp. strains in the post-genome era. Mol. Plant Microbe Interact. *24*, 1310–1316.

Bertalan, M., Albano, R., de Padua, V., Rouws, L., Rojas, C., Hemerly, A., Teixeira, K., Schwab, S., Araujo, J., Oliveira, A., *et al.* (2009). Complete genome sequence of the sugarcane nitrogen-fixing endophyte *Gluconacetobacter diazotrophicus* Pal5. BMC Genomics *10*, 450.

Bove, J.M., and Garnier, M. (2002). Phloem-and xylem-restricted plant pathogenic bacteria. Plant Science *163*, 1083–1098.

Bronstein, J.L., Alarcon, R., and Geber, M. (2006). The evolution of plant-insect mutualisms. New Phytol. *172*, 412–428.

Brundrett, M. (2004). Diversity and classification of mycorrhizal associations. Biol. Rev. *79*, 473–495.

Brundrett, M.C. (2002). Coevolution of roots and mycorrhizas of land plants. New Phytol. *154*, 275–304.

Bucher, M., Wegmueller, S., and Drissner, D. (2009). Chasing the structures of small molecules in arbuscular mycorrhizal signaling. Curr. Opin. Plant Biol. *12*, 500–507.

Compant, S., Clement, C., and Sessitsch, A. (2010). Plant growth-promoting bacteria in the rhizo- and endosphere of plants: Their role, colonization, mechanisms involved and prospects for utilization. Soil Biol. Biochem. *42*, 669–678.

Denison, R.F., and Kiers, E.T. (2011). Life histories of symbiotic rhizobia and mycorrhizal fungi. Curr. Biol. *21*, R775-R785.
Denison, R.F., Bledsoe, C., Kahn, M., O'Gara, F., Simms, E.L., and Thomashow, L.S. (2003). Cooperation in the rhizosphere and the "free rider" problem. Ecology *84*, 838–845.
Dennis, P., Hirsch, P., Smith, S., Taylor, R., Valsami-Jones, E., and Miller, A. (2009). Linking rhizoplane pH and bacterial density at the microhabitat scale. J. Microbiol. Methods *76*, 101–104.
Dennis, P.G., Miller, A.J., and Hirsch, P.R. (2010). Are root exudates more important than other sources of rhizodeposits in structuring rhizosphere bacterial communities? FEMS Microbiol. Ecol. *72*, 313–327.
Douglas, A.E. (2008). Conflict, cheats and the persistence of symbioses. New Phytol. *177*, 849–858.
Eaton, C.J., Cox, M.P., and Scott, B. (2011). What triggers grass endophytes to switch from mutualism to pathogenism? Plant Sci. *180*, 190–195.
Finlay, R.D. (2008). Ecological aspects of mycorrhizal symbiosis: with special emphasis on the functional diversity of interactions involving the extraradical mycelium. J. Exp. Bot. *59*, 1115–1126.
Fleming, T.H., and Muchhala, N. (2008). Nectar-feeding bird and bat niches in two worlds: pantropical comparisons of vertebrate pollination systems. J. Biogeogr. *35*, 764–780.
Franche, C., Lindstrom, K., and Elmerich, C. (2009). Nitrogen-fixing bacteria associated with leguminous and non-leguminous plants. Plant Soil *321*, 35–59.
Gans, J., Wolinsky, M., and Dunbar, J. (2005). Computational improvements reveal great bacterial diversity and high metal toxicity in soil. Science *309*, 1387–1390.
Gough, C., and Cullimore, J. (2011). Lipo-chitooligosaccharide signaling in endosymbiotic plant–microbe interactions. Mol. Plant Microbe Interact. *24*, 867–878.
Gyaneshwar, P., James, E.K., Reddy, P.M., and Ladha, J.K. (2002). *Herbaspirillum* colonization increases growth and nitrogen accumulation in aluminium-tolerant rice varieties. New Phytol. *154*, 131–145.
Hammond-Kosack, K., and Jones, J.D.G. (2000). Responses to plant pathogens. In Biochemistry and Molecular Biology of Plants, Buchanan, B.B., Gruissem, W., and Jones, R.L., eds (American Society of Plant Biology, Rockville, MD), pp.1102–1156.
Herridge, D.F., Peoples, M.B., and Boddey, R.M. (2008). Global inputs of biological nitrogen fixation in agricultural systems. Plant Soil *311*, 1–18.
Hirsch, P.R. (1996). Population dynamics of indigenous and genetically modified rhizobia in the field. New Phytol. *133*, 159–171.
Hurek, T., Handley, L.L., Reinhold-Hurek, B., and Piche, Y. (2002). *Azoarcus* grass endophytes contribute fixed nitrogen to the plant in an unculturable state. Mol. Plant Microbe Interact. *15*, 233–242.
Jones, J.D.G., and Dangl, J.L. (2006). The plant immune system. Nature *444*, 323–329.
Keeling, P.J. (2010). The endosymbiotic origin, diversification and fate of plastids. Philos. Trans. R. Soc. B-Biol. Sci. *365*, 729–748.
Krause, A., Ramakumar, A., Bartels, D., Battistoni, F., Bekel, T., Boch, J., Boehm, M., Friedrich, F., Hurek, T., Krause, L., *et al.* (2006). Complete genome of the mutualistic, N-2-fixing grass endophyte *Azoarcus sp* strain BH72. Nat. Biotechnol. *24*, 1385–1391.
Lindow, S.E., and Brandl, M.T. (2003). Microbiology of the phyllosphere. Appl. Environ. Microbiol. *69*, 1875–1883.
Lugtenberg, B., and Kamilova, F. (2009). Plant-growth-promoting rhizobacteria. Annu. Rev. Microbiol. *63*, 541–556.
Lynch, J.M., and Whipps, J.M. (1990). Substrate flow in the rhizosphere. Plant Soil *129*, 1–10.
Masson-Boivin, C., Giraud, E., Perret, X., and Batut, J. (2009). Establishing nitrogen-fixing symbiosis with legumes: how many rhizobium recipes? Trends Microbiol. *17*, 458–466.
Mendes, R., Kruijt, M., de Bruijn, I., Dekkers, E., van der Voort, M., Schneider, J.H.M., Piceno, Y.M., DeSantis, T.Z., Andersen, G.L., Bakker, P.A., *et al.* (2011). Deciphering the rhizosphere microbiome for disease-suppressive bacteria. Science *332*, 1097–1100.
Millet, Y.A., Danna, C.H., Clay, N.K., Songnuan, W., Simon, M.D., Werck-Reichhart, D., and Ausubel, F.M. (2010). Innate immune responses activated in arabidopsis roots by microbe-associated molecular patterns. Plant Cell 22, 973–990.
Minchin, F.R., and Pate, J.S. (1973). Carbon balance of a legume and functional economy of its root nodules. J. Exp. Bot. *24*, 259–271.
Oldroyd, G.E.D., Harrison, M.J., and Paszkowski, U. (2009). Reprogramming plant cells for endosymbiosis. Science *324*, 753–754.
Parniske, M. (2008). Arbuscular mycorrhiza: the mother of plant root endosymbioses. Nat. Rev. Microbiol. *6*, 763–775.

Paszkowski, U. (2006). Mutualism and parasitism: the yin and yang of plant symbioses. Curr. Opin. Plant Biol. *9*, 364–370.

Pedrosa, F.O., Monteiro, R.A., Wassem, R., Cruz, L.M., Ayub, R.A., Colauto, N.B., Fernandez, M.A., Fungaro, M.H.P., Grisard, E.C., Hungria, M., *et al.* (2011). Genome of *Herbaspirillum seropedicae* strain SmR1, a specialized diazotrophic endophyte of tropical grasses. Plos Genetics *7*.

Rasmussen, H.N. (2002). Recent developments in the study of orchid mycorrhiza. Plant Soil *244*, 149–163.

Reid, D.E., Ferguson, B.J., Hayashi, S., Lin, Y.-H., and Gresshoff, P.M. (2011). Molecular mechanisms controlling legume autoregulation of nodulation. Ann. Bot. *108*, 789–795.

Reinhold-Hurek, B., and Hurek, T. (2011). Living inside plants: bacterial endophytes. Curr. Opin. Plant Biol. *14*, 435–443.

Rosenblueth, M., and Martinez-Romero, E. (2006). Bacterial endophytes and their interactions with hosts. Mol. Plant Microbe Interact. *19*, 827–837.

Saikkonen, K., Wali, P., Helander, M., and Faeth, S.H. (2004). Evolution of endophyte-plant symbioses. Trends Plant Sci. *9*, 275–280.

Schardl, C.L., Leuchtmann, A., and Spiering, M.J. (2004). Symbioses of grasses with seedborne fungal endophytes. Annu. Rev. Plant Biol. *55*, 315–340.

Sevilla, M., Burris, R.H., Gunapala, N., and Kennedy, C. (2001). Comparison of benefit to sugarcane plant growth and N-15(2) incorporation following inoculation of sterile plants with *Acetobacter diazotrophicus* wild-type and Nif(-) mutant strains. Mol. Plant Microbe Interact. *14*, 358–366.

Smith, M.J., Adams, M.J., and Ward, E. (2011). Evidence that *Polymyxa* species may infect *Arabidopsis thaliana*. FEMS Microbiol. Lett. *318*, 35–40.

Smith, S.E., Smith, F.A., and Jakobsen, I. (2003). Mycorrhizal fungi can dominate phosphate supply to plants irrespective of growth responses. Plant Physiol. *133*, 16–20.

Spoel, S.H., and Dong, X. (2012). How do plants achieve immunity? Defence without specialized immune cells. Nat. Rev. Immunol. *12*, 89–100.

Tani, C., Ohtomo, R., Osaki, M., Kuga, Y., and Ezawa, T. (2009). ATP-dependent but proton gradient-independent polyphosphate-synthesizing activity in extraradical hyphae of an arbuscular mycorrhizal fungus. Appl. Environ. Microbiol. *75*, 7044–7050.

Tellstroem, V., Usadel, B., Thimm, O., Stitt, M., Kuester, H., and Niehaus, K. (2007). The lipopolysaccharide of *Sinorhizobium meliloti* suppresses defense-associated gene expression in cell cultures of the host plant *Medicago truncatula*. Plant Physiol. *143*, 825–837.

Treseder, K.K., and Allen, M.F. (2002). Direct nitrogen and phosphorus limitation of arbuscular mycorrhizal fungi: a model and field test. New Phytol. *155*, 507–515.

Uma, B., Rani, T.S., and Podile, A.R. (2011). Warriors at the gate that never sleep: Non-host resistance in plants. J. Plant Physiol. *168*, 2141–2152.

Vessey, J.K., Pawlowski, K., and Bergman, B. (2004). Root-based N_2-fixing symbioses: Legumes, actinorhizal plants, *Parasponia* sp and cycads. Plant Soil *266*, 205–230.

Zamioudis, C., and Pieterse, C.M.J. (2011). Modulation of host immunity by beneficial microbes. Mol. Plant Microbe Interact. *25*, 139–150.

A Bird's Eye View of Microbial Community Dynamics

4

Zhanshan (Sam) Ma, Jiawei Geng, Zaid Abdo and Larry J. Forney

Abstract

Microbial community dynamics is one of the most important central themes of microbial community ecology, which seems to be experiencing its first golden era thanks to the rapidly expanding datasets derived using metagenomic and other '-omics' methods in microbial biology. For example, much of the ongoing NIH-HMP (Human Microbiome Project) focus has been centred on the dynamics of human microbiome communities. Microbial ecologists are beginning to actively draw upon ecological theories from macro ecology to study microbial communities. In this article, we present a brief review on several selected topics of ecological theories that are most relevant to community dynamics, including the diversity–stability paradigm, intermediate disturbance hypothesis (IDH), species area/time curves (SAT), species abundance distribution (SAD), and neutral community theory. In perspective, we suggest that the study of microbial community dynamics can not only benefit from applying ecological theories originally developed in macroecology, but also contribute to the development and testing of new ecological theories. These bidirectional interactions are of critical importance to the flourishing of theoretical microbial ecology, and studies of microbial community dynamics offers tremendous opportunities for these intellectual exchanges to occur.

Introduction

The dynamics of ecological communities can be defined as temporal and spatial changes in community structure or function. Ideally, community dynamics would be studied from the perspectives of both structure and function but most often this has not been the case. Instead, studies of community dynamics often centre on the analysis of only community diversity in time and space. Some of the most commonly studied properties of community dynamics include: diversity (evenness and richness), spatial and temporal changes in diversity (often termed β-diversity and γ-diversity), and diversity–stability–disturbance relationships. From these some of the most prominent concepts in community ecology have been explored, most notably, species–area–time (SAT) curves, species abundance distributions (SAD), and the intermediate disturbance hypothesis (IDH). These and related topics are of growing interest to microbial ecologists thanks to recent advances in metagenomic studies of microbial communities. Other topics such as food webs, ecological networks, equilibrium versus non-equilibrium community dynamics, and the roles of cooperation and communication in microbial communities are also of fundamental importance to the study of community dynamics but beyond the scope of this chapter.

Diversity, stability, and disturbance: three fundamentals of community dynamics

Ecological theories are often imprecise and lack the rigor of those found in other scientific disciplines. Biological systems have properties typical of complex systems including self-organization, dominance of non-linearity, and unpredictability of environmental change. These confound efforts to rigorously define ecological theories. But perhaps the more fundamental problem is that 'evolution hates definitions' (Bradbury and Vehrencamp, 1998) – the reality that evolution seems to always create exceptions to otherwise general rules. Thus we are left with concepts that form the foundation of our understanding and guide experimentation, but that lack the universality that some might desire.

Community diversity

Community diversity can be defined on the basis of *species abundance* and *species richness*. The first encompasses the *evenness* of species while the second reflects the number of species in the community. Theoretically, all information about community evenness can be captured by a multivariate probability distribution of relative species abundance of all species in a community, or SAD (species abundance distribution) as defined by

$$\text{SAD} = P(S_1 = n_1, S_2 = n_2,, S_R = n_R) \tag{4.1}$$

where R is the number of species or *richness* (R). A SAD is more informative and flexible than diversity indices and multivariate analysis for diversity research. However, it is limited by its *static* nature since only a snapshot of species distribution is captured.

The computational complexity of determining multivariate probability distributions was largely responsible for its loss of favour to diversity indices that began to thrive in the 1960s. However, diversity indices are limited in two important ways. First, the indices are calculated solely on the basis of *observed* richness and evenness, and the actual *underlying* richness and evenness is too seldom taken into account. This is especially problematic in microbial communities that are densely populated, and typically dominated by few species with a 'long tailed distribution' of rare species that are usually not sampled or enumerated. Second, as noted by Southwood and Henderson (2000) there 'has been an 'explosive speciation' of diversity indices, which initially brought confusion to the subject; in addition, the ubiquitous nature of some relationships and the apparent constancy of certain numerical values have added a measure of mystique.'

Shannon and Simpson indices

Two of the most widely used diversity indices are the *Shannon* index and the *Simpson* index. The *Shannon entropy index* is computed from:

$$H_d = -\sum p_i \ln(p_i) \tag{4.2}$$

where p_i is the proportion of individuals of the ith species in the community. The Shannon index also has a companion *evenness* index, which is often computed from:

$$H_e = H_d/H_{d\max} = H_d/\ln(S) \tag{4.3}$$

where H_e is the *Shannon evenness index* and S is the number of species (S) in the community. It is claimed that H_d captures both aspects of community diversity and H_e captures the evenness. There have been several attempts to improve the index that have focused on devising an evenness index that is independent of species richness (e.g. see comprehensive discussions in Magurran, 2004, and Southwood and Henderson, 2000). Magurran's (2004) monograph summarized the properties of those indices, including their perceived properties that are often ignored by most applied ecologists. Rather than attempting to figure out possible differences revealed by different diversity indices it is probably more important to consistently use a single index so that the results computed from multiple times or locations can be compared.

The Shannon index is often hailed as a better index owing to its apparent connection with Shannon's entropy in his information theory. In fact, many probability distributions including Weibull distribution has a defined entropy that we have found has wide applicability for modelling diversity and richness (Ma, 1991a, 2011; Ma, unpublished results). The entropy index has the following form:

$$H=\gamma\left(1-\frac{1}{\beta}\right)+\ln\left(\frac{1}{\beta\lambda}\right)+1 \tag{4.4}$$

where γ is the Euler–Mascheroni constant, and equal to 0.57721566490153286065 ... β and λ are the shape and scale parameters of a Weibull distribution. In our opinion, the probability distribution itself should be preferred over the composite entropy index to further characterize the dynamics of diversity.

Community stability

Grimm and Wisse (1997) catalogued 163 definitions of 70 different stability concepts, as well as various models for those definitions. They were able to partition the definitions into three broad categories: *resistance, resilience,* and *persistence*: (i) resistance (or constancy) – staying essentially unchanged over time; (ii) resilience – returning to the reference state (or dynamic trajectory) after a temporary disturbance; and (iii) persistence – persistence of an ecological state through time. Arguably the most widely used approach to stability modelling is still the differential equation approach pioneered by May (1973) and more recently expanded by Mueller and Joshi (2000). Although we subscribe to Grimm and Wisse's (1997) categorization of stability concepts it is often not feasible to parameterize the *deterministic* stability definitions with practically available data and most have only been qualitatively analysed by examining the stability properties of those differential equations.

Disturbance

A disturbance may be defined as any relatively discrete event in time that disrupts an ecosystem, community, or population structure and simultaneously changes resources, substrate availability, or the physical environment (Putman, 1994). Alternatively, a disturbance can be defined as a discrete, punctuated killing, displacement, or damage to one or more populations in a community that directly or indirectly creates opportunities for new individuals to become established (Sousa, 1984; Putman, 1994). These definitions are similar, but place emphasis on different targets of disturbance. It should be noted that both biological and abiotic (physical) processes may act as agents of disturbance, and the boundary between the biological community and environment is blurred and interactive. For example, in the case of the human microbiome a microbial community has reciprocal interactions with its human host (environment).

Intermediate disturbance hypothesis

The study of disturbances has two important ramifications in community ecology: the maintenance of community diversity, and the complex relationship of diversity and community stability. There are two major hypotheses regarding the relationship between diversity and disturbance. One is the *disturbance frequency hypothesis,* which argues that diversity is maintained because disturbance recurs more frequently than the time required for competitive exclusion (Huston, 1979). Another one is the *intermediate disturbance hypothesis* (IDH) that asserts species richness will be greatest in communities that experience some intermediate level of disturbance (Connell, 1978). Obviously, this statement is vague because quantifying 'intermediate' levels of disturbance can be challenging. Roxburgh *et al.* (2004) distinguished three models, namely a *spatial within-patch* model, a *spatial between-patch* model, and a *purely temporal* model that all generate similar patterns of coexistence in response to intermediate disturbances.

At least three, quite distinct, mechanisms can lead to species coexistence: the *storage effect, niche partitioning,* and *relative non-linearity* (Chesson, 2000a,b). The storage effect refers to subadditive or buffered population growth. This occurs when an inferior competitor expands rapidly following a disturbance when competitive pressure may be low. At the end of a disturbance cycle competition is intensified and the population growth rate of all community members is reduced. However, the gains made by weak competitors during the most favourable periods of growth are not negated. This 'storage effect' is often evident when the weaker competitor 'stores' members of its population in patches of a post-disturbance landscape that are protected from superior competitors in a spatial between-patch model. The niche partitioning is more about competition and dispersal *within a patch,* and does not depend on fluctuating environmental conditions (Roxburgh *et al.,* 2004). In other words, intermediate disturbance facilitates niche partitioning and promotes species diversity. In contrast, relative non-linearity is a temporal mechanism that causes species coexistence. This results when the response of a superior competitor to temporal variations in resource availability (and hence competition) is highly non-linear relative to that of an inferior competitor (Roxburgh *et al.,* 2004).

A simplified view of the diversity–stability paradigm

The relationships between diversity, stability, and disturbance have been debated for many years (e.g. Green *et al.,* 2005; Neutel *et al.,* 2007). To some extent, the issue is similar to the familiar conundrum of the casual relationship between a chicken and egg (Green *et al.,* 2006). In the context of complexity science it is stability that allows diversity to develop, and a consequence is the emergence of increased connectivity. These intraspecific and interspecific interactions take many forms and occur in a hierarchical network with complex positive (e.g. mutualism) and negative (e.g. predation) feedbacks, and non-linear processes (e.g. chaotic population dynamics). Thus, the resulting rise of connectivity may stabilize or destabilize a system depending on the underlying interaction mechanisms. Superimposed on this are disturbances of various types and intensities, and the evolutionary history of the community. Collectively these all affect the outcome of interspecies interactions and the emergent properties of diversity and stability.

Species–area curves and island biogeography: implications for modelling species richness and turnover

Species–area curves

Area has long been known to have a pervasive and powerful influence on species richness in macro-ecology (Arrhenius, 1921) and species–area relationships (SPAR) can often be described by the following power function:

$$S = aA^b \tag{4.5}$$

where A is area, S is the number of species, a and b are parameters. Preston (1962) speculated that space and time (especially evolutionary time) should have equivalent effects in terms of the accumulation of species richness, and this has become known as Preston's Ergodic conjecture. The species–time relationship (STR) has the exactly same form as SPAR,

$$S = cT^d \tag{4.6}$$

where T is time, S is the number of species, c and d are parameters. Finally, there is a species–area–time (SAT) curve proposed by Rosenzweig (1998):

$$S_{\mathrm{TA}} = S(t)A^{\mathrm{b}} \tag{4.7}$$

In the past decade researchers have worked to separate the effects of sampling, as well as ecological and evolutionary processes on species accumulation in space and time. One of the most important advances is the attempt to separate sampling effects from real ecological and evolutionary processes that may be responsible for the species accumulation over space and time. This sampling effect, which is also known as 'rarefaction effect', 'random placement effect', or 'passive sampling', refers to the phenomenon that more species are accumulated when more individuals or samples are collected even though the sampling universe is fixed. Obviously, any meaningful ecological/evolutionary interpretation attached to SPAR, STR, or SAT must first reject the sampling effect.

The sampling effect for SPAR was already studied in the early days of SPAR (Arrhenius, 1921). White *et al.* (2004, 2010) studied the sampling effect for STR by extending the Coleman (1981) model. Recently, McGlinn and Palmer (2009) built the first sampling effect model for the SAT model based on first principles. It is a stochastic model by assuming that species interactions are neutral (Hubbell, 2001). In other words, the species accumulation process is solely determined by the distribution of the relative abundances in the sampling universe of the species pool, and environment or the intrinsic difference among species does make a difference. Mathematically, the assumption is expressed as follows:

$$J = A + AR(T - 1) \tag{4.8}$$

where J is the number of individuals in the local community and it is held constant over time owing to the assumption that neutral community experiences zero-sum dynamics. A is the sampling area. R is the probability that an individual in the local community will be replaced by individuals randomly selected from the species pool during one time unit. T is the time unit. The expected number of species for J randomly sampled individuals is then:

$$E\left[S(J|R,p)\right]=\sum_{i=1}^{S_p}\left[1-(1-p_i)^J\right]=S_p-\sum_{i=1}^{S_p}(1-p_i)^J \tag{4.9}$$

where S_p is the total number of species in the species pool, and p is a vector representing the relative abundance distribution of the species pool, i.e. $p = [p_1, p_2, \ldots, p_i, \ldots p_{Sp}]$. The conclusion McGlinn and Palmer (2009) drew is that unless individual replacement rate is one, time and space are not symmetrical in terms of species accumulation. Furthermore, both positive and negative interactions between space and time are possible in their sampling model. This implies that non-zero interactions between space and time are not necessarily the evidence of ecological processes.

Harte *et al.* (2008, 2009) proposed a model for SPAR in the following form:

$$S(A)=S_0\sum_{n=1}^{N_0}\left[1-P(0|n,A,A_0)\right]\Phi(n|S_0,N_0) \tag{4.10}$$

where A_0 is the area at some specified largest space scale, and S_0 and N_0 are the corresponding total number of species and individuals, respectively. A is the area of an arbitrarily selected cell nested within the area A_0, and $S(A)$ is the expected number of species in A. The first item within the summation symbol, $[1 - P(0|n, A, A_0)]$, is the probability that a species with n individuals in A_0 is present in such a cell of area A. It is a *spatial abundance distribution*. The second item $\Phi(n|S_0, N_0)$ is the *species abundance distribution*, i.e. the probability that a species picked at random from the species pool in A_0 has n individuals. The authors further applied the *maximum entropy principle* in non-equilibrium statistical mechanics to derive the estimation procedure for $S(A)$. Equation 4.10 allows the estimation of $S(A)$ both down-scaling and up-scaling A_0.

Scheiner *et al.* (2003) emphasized the need to pay attention to the differences in the methods for constructing SPAR. They identified six types of SPAR, depending on (i) the patterns of areas sampled (nested, contiguous, non-contiguous, or island), (ii) whether successfully larger areas are spatially explicit or not; (iii) whether the curve is developed from single values or mean values.

Rosindell and Cornell's (2009) simulation study with spatially explicit neutral models display three typical phases of SPAR, i.e. (i) sampling phase at small local sampling areas, where each individual is a new species; (ii) Arrhenius power law phase, $S \propto A^Z$; and (iii) a third 'continental' phase where regions that are separated by large distances will contain non-overlapping sets of species, S A. They concluded that neutral models with biologically reasonable dispersal and speciation parameters can lead to empirically realistic SPAR curves.

SPAR curves are often associated with estimates of species richness in which one is concerned with the accumulation of new species as the sampling area or sampling effort is increased. For this reason, Ugland *et al.* (2003) distinguished between a 'species-area curve' and a 'species accumulation curve'. The former describes the number of species in areas of different sizes irrespective of the identity of the species within the areas; the latter describes the accumulation rates of new species over the sampled area and depends on species identity. Although this distinction in terminology has not been followed in general in the literature, it is of crucial importance for properly estimating species richness. Another contribution Ugland *et al.* (2003) made is to derive exact analytical expectation and variance of the species accumulation curve in all subsets of samples from a given area, from which a new total species curve (T–S curve) is obtained. It is the *T–S* curve that can be extrapolated to estimate the total number of species in the area investigated. There are three major hypotheses for why SPAR exist: (i) the *habitat diversity hypothesis* that assumes a positive correlation among habitat size, habitat heterogeneity, and species diversity; (ii) the previously mentioned passive sampling or *random placement hypothesis*; and (iii) the *dynamic equilibrium hypothesis*. The most influential is the latter, which lies at the core of MacArthur and Wilson's (1967) *island biogeography theory*.

Island biogeography

The theory of island biogeography posits that the species diversity on an island is determined by two processes: *local extinction* and *immigration* from species pools in mainland habitats (MacArthur and Wilson, 1967; Wu and Vankat 1995). For a given area and distance to the mainland the extinction and immigration rates should monotonically increase and decrease, respectively, as the number of species on the island increases. When the two rates are equal, the species diversity on the island becomes relatively constant, but the composition of species continues to change owing to concurrent extinction and immigration processes. In other words, the biotic diversity on the island reaches a dynamic equilibrium. The dynamic change of species composition is referred to as *turnover*, and the turnover rate at equilibrium is theoretically equal to the immigration or extinction rate. Island biogeography theory also predicts the so-termed *area effect*: as the area increases, the species population size becomes larger and the probability of extinction decreases (because large populations are less prone to extinction). The area effect therefore dictates the positive correlation between island size and species richness. As with many ecological theories, the assumptions used in island biogeography theory may be too simplified for many natural communities, and the strict dynamic equilibrium predicted by the theory may not exist in complex communities.

Species richness and turnover

It may be seen from the above review, that SPAR and island biogeography theory are at the heart of species richness dynamics. It should also be recognized that the closely associated concept of species turnover might reveal community changes even when the richness (diversity) stays relatively constant and when the community is in dynamic equilibrium. Russell (1998) listed three properties of turnover: (i) turnover exists in all biological systems; (ii) turnover can be modelled in a similar way over space and time; (iii) turnover may be measured in a similar way over different scales. One simple fact should justify why it is important to study turnover besides studying community diversity and dominance: change of diversity (richness) implies occurrence of turnover, but diversity may fail to reflect the occurrence of turnover. In other words, turnover events may or may not lead to change in diversity (richness). Russell (1998) defined turnover simply as the difference in composition between two community censuses.

$$\text{Turnover} = \frac{(\text{Number of appearances}) + (\text{Number of disappearances})}{(\text{Total number of species present}) \times (\text{Census interval})} \quad (4.11)$$

Species abundance distributions and community neutral theory

Species abundance distributions

Biological communities are assemblages of multiple species that share a common habitat. Hence the proportion (distribution) of the member species' abundances is a key variable in microbial community ecology. In the very early days of ecology it attracted the interests of eminent statisticians such as Ronald A. Fisher (1943), who developed the *log-series distribution* model and α-diversity, which is still widely used today. Another legendary figure that made significant contribution to this approach was Frank Preston (1948, 1962), who discovered the wide applicability of the *lognormal distribution* in describing SAD. The original Fisher log-series model is indeed an infinite series of the expected number of species (E_r) with r individuals:

$$E_r = \frac{\alpha X^r}{r}, \quad (4.12)$$

where $\alpha > 0$ and $0 < X < 1$ are two defining parameters. The expected number of species in the sample is:

$$E(S) = -\alpha \log(1 - X) \quad (4.13)$$

and the expected number of individuals in the sample is:

$$E(N) = \alpha X / (1 - X) \quad (4.14)$$

The maximum likelihood estimate of α is the solution to the following function:

$$S = \alpha \log\left(1 + \frac{N}{\alpha}\right) \quad (4.15)$$

The variance of α is equal to:

$$\text{var}(\alpha) = -\alpha / \log(1 - X) \; (4.16)$$

Taylor and Kemp (1976) found that α can be an effective diversity index for habitat discrimination that behaves more predictably and consistent than the more commonly used Shannon and Simpson diversity indices.

Mathematically, SAD may be defined as a vector of the abundances of all species in a community (McGill *et al.*, 2007, 2010). Graphically, what makes SAD a universal law in community ecology is the classic hyperbolic J-shaped curve, which is also known as lazy

J-curve or hollow-curve demonstrated by SAD when it is plotted as a histogram of percentage or proportion of species on the *y*-axis versus species abundance on an arithmetic *x*-axis.

Neutral community theory

No discussion of SAD would be complete without mention of neutral community theory (Hubbell, 2001), which is arguably the most significant advance in community ecology in the last decade. The primary result of the neutral community theory, also known as the unified neutral theory of biodiversity and biogeography, is a synthetic explanation of species abundance in space, time and across scales (Chave, 2004). Furthermore, the neutral theory offers a common thread to unify some of the most important ecological paradigms including SAD, SPAR, and island biogeography. The debate surrounding its validity is fascinating and much of the theory seems to have survived the rigorous scrutiny as ecologists sought to reconcile neutral theory with niche-based theories of community assembly.

A detailed discussion on neutral theory is beyond the scope of this chapter. Here, we only briefly mention some original aspects of neutrality assumptions and the resulting theory. Prior to Hubbell's (2001) formal proposal of his community neutral theory, partial ideas on neutral theory in community ecology had already been attempted by several ecologists including MacArthur and Wilson (1967), Watterson (1974), Caswell (1976) and Leigh *et al.* (1993). These neutrality ideas apparently were inspired by Kimura's (1964, 1986, 1991) neutral theory of molecular evolution. Kimura argued that most evolutionary changes at the molecular level are neutral and do not cause changes in an organism's fitness. A typical fitness landscape is then flat according to this theory. The neutral theory stresses the possibility of extensive genetic divergence by stochastic factors in the absence of deterministic forces of selection. Similarly, in a community, when the species differences in a community do not lead to ecologically competitive advantages, the community is said to be neutral and the species are considered as functionally equivalent (Alonso *et al.*, 2006). While the species interactions are assumed to be neutral, the demographic stochasticity and migration are deemed responsible for much of the ecological divergence in community. According to Alonso *et al.* (2006), the originality of Hubbell's (2001) neutral theory lies in the combination of the following aspects: (i) the assumption of functionally equivalent interacting species; (ii) it is an individual-based stochastic dynamics theory; and (iii) it is a dispersal-limited sampling theory.

Indeed, Hubbell's (2001) original formulation was based on two biological assumptions. The first is the neutrality, equivalence, or symmetry assumption: different individuals from different species belonging to the same functionally uniform ecological community are controlled by similar birth, death and dispersal rates. The second assumption is that ecological communities are saturated, which leads to the zero-sum dynamics among individuals. The neutrality assumption raises two critical issues. First, is the assumption of neutrality realistic? And second, if neutrality is not realistic, then what mechanisms are responsible for the observed departure from neutrality.

The neutral community theory possesses three features: (i) it defines the stochastic dynamics of species from their origin to extinction with its fundamental number of speciation; (ii) it has a spatial formulation with its fundamental number of dispersal; (iii) it considers the dynamics of discrete individuals by modelling demographic stochasticity. The three features enable quantitative predictions of very general ecological patterns such as

SAD, SPAR, the spatial and temporal turnover of species, the distribution of species ranges, and the relationship between species range and abundance (Alonso *et al.*, 2006).

As reviewed by Chave (2004), Hubbell's (2001) neutral theory displays its heuristic power by successfully predicting many basic patterns of biodiversity. But it appears to have little empirical evidence in general to support the strict assumption of equivalence that defines neutrality, i.e. the equivalence among individuals. A weaker assumption of neutrality may hold for stable communities, i.e. the equivalence of average fitness among species, which is congruent with other theories of species coexistence that satisfy the weaker equivalence assumption, such as niche differentiation theory.

In population genetics, critics of Kimura's (1964, 1986, 1991) neutral theory of molecular evolution sometimes dismiss the theory as a mere null hypothesis, or a theory of 'no effect'. As a response to the critics, there is the so-termed 'near-neutrality theory.' The near-neutrality theory is considered as the leading edge corollary of the original neutral theory, and it offers a more powerful explanation for the observed evolutionary patterns (Hughes, 2008). For example, with near-neutrality theory it is argued that slightly deleterious mutations will ultimately be eliminated by positively selected compensatory changes so long as the effective population size is large enough. This is consistent with Kimura's original arguments: although genetic drift and purifying selection predominate at the molecular level, positive selection indeed might occur, although it is relatively rare (Hughes, 2008). In our opinion, the transition from neutral theory to near neutral theory in molecular evolution should inspire us to further our efforts in unifying neutral community theory with niche-based assembling theory. The dichotomy of genetic drift versus positive selection force in molecular evolution seems to mirror the counterpart of ecological drift (dispersal-based community assembly) versus non-neutral interactions (niche-partition community assembly) in community dynamics.

In arguing for the advantages of neutral and near-neutrality theory over 'selectionism lite', Hughes (2008) cited the philosopher Daniel Dennett's (1995) claim that Darwin's key (and 'dangerous') insight was that evolution is an 'algorithmic process,' which implies that the evolution is essentially deterministic. Hughes (2008) further compared the neutrality idea with Heisenberg's well-known 'uncertainty principle' in physics and Gödel's 'proof of incompleteness' in mathematics. In our opinion, over emphasizing the difference between selectionist and neutralist views may be unnecessary and not necessarily beneficial for understanding evolution because the 'neutrality' and 'positive selection' views are not mutually exclusive and there is room for both theories. We hope ecologists will not repeat the evolutionary version of the 'selectionist–neutralist controversy' of the 1970s. Actually, recent research (e.g. Ma, 2011) in evolutionary computing suggests that 'algorithmic process' can contain 'neutral space' in the fitness landscape of evolutionary computing, which is inspired by the fitness landscape in population genetics. The 'algorithmic process' in evolutionary computing is both deterministic and stochastic. On the one hand, it is deterministic from the perspective that the computed results are generally deterministic (otherwise, we cannot rely on the results from evolutionary computing). On the other hand, the search for a solution does include neutral space in the fitness landscape, and the search process can be stochastic.

Woodcock *et al.* (2007) presented what they claimed then to be the most convincing evidence for neutral community assembling in both microbial and macro-ecology with 27 waterborne bacterial communities housed in bark-lined tree holes of 27 beech trees. They tested two hypotheses derived from neutral community theory. First, do the samples derive from the same SAD distribution, i.e. whether or not the same structuring force shaped the tree-hole bacterial communities? The data rejected the hypothesis. Second, do immigrants from a single source meta-community seed neutrally assembled communities with a constant immigration rate? In Hubbell's (2001) model, it is assumed that the SAD in the source community is described by a log-series distribution with a single parameter θ, i.e. *the fundamental number of biodiversity*, with the assumption that the parameter indexes the overall biodiversity. In local communities, where zero-sum replacement is enforced because the communities are saturated with individuals, a dying individual is either replaced with probability m by an immigrant drawn at random from the source meta-community, or by reproduction from the local community with the probability of $1 - m$, which depends on the SAD of local community because the local species abundance obviously affects which species will contribute the next offspring to the community. The SAD in turn is also influenced by the total number of individuals in the community, N_t. Overall, the shape of SAD for a neutrally assembled community is determined by three parameters: θ, m and N_t. In the study conducted by Woodcook *et al.* (2007), the parameter N_t was estimated by the density of organisms and the volume of the tree hole, and the other two parameters were treated as free parameters and adjusted to give the best least-square of error fit to their datasets. It turned out that among 29 datasets, the neutral theory model was rejected in only two datasets. In the other 27 datasets, there was no evidence to reject the neutral model. Based on these results, Woodcock *et al.* (2007) suggested that at least some bacterial communities are dispersal limited, and they challenged the views that global dispersal of microorganisms prevents them from having biogeography and that microbial populations are large enough to save them from local stochastic extinctions.

Space is often said to be the final frontier of ecology. Given that neutral theory is considered as a dispersal-limited sampling theory, it should be natural and interesting to test the neutral theory in the context of spatial explicit versus spatial implicit models. In a recent study, Etienne and Rosindell (2011) did just this by approximating a spatially explicit model with a spatially implicit model because it is often too difficult to incorporate a spatially explicit model into neutral theory model owing to the analytical complexity of the spatially explicit model. They concluded that current spatially explicit neutral models possess limited descriptive power, and that more advanced spatially explicit models should be developed to analyse the spatial limitation of current neutral theory models.

Perspective

Microbial ecologists have historically been limited to studies of autecology and population ecology because of the seemingly insurmountable challenges faced in studies of microbial community composition and dynamics. However, the development of metagenomic profiling methods to study microbial communities that obviate the need for cultivation have extended our horizons. Now the challenge is for microbial community ecologists to articulate objectives for the emerging field of theoretical microbial ecology in the metagenomic and post-metagenomic era. We suggest two complementary objectives: (i) to test existing ecological theories using data available from the metagenomic and other '-omics' methods now used in molecular microbial biology; and (ii) develop and explore ecological theories that seem particularly relevant to microbial community ecology such as food web and network analyses, equilibrium versus non-equilibrium dynamics, cooperation, and communication (e.g. Zengler, 2009; Robinson *et al.*, 2010). From our perspective it seems that microbial community ecology is now entering its golden era. This is exemplified by the increasingly evident role that complex communities of the human microbiome have on health and risk to disease. And so we contend that the emerging new role of microbial ecology should not only enrich ecological theories but also help to push ecology closer to centre stage in understanding ways to promote environmental sustainability and the health of plants, animals and human beings.

Acknowledgements

This work was supported by grants UO1 AI070921 from the National Institute of Allergy and Infectious Diseases and UH2 AI083264 from the Human Genome Research Institute of the National Institutes of Health.

References

Alonso, D., Etienne, R.S., and McKane A.J. (2006). The merits of neutral theory. Trends Ecol. Evol. *21*, 451–457.

Arrhenius, O. (1921). Species and area. J. Ecol. *9*, 95–99.

Bradbury, J.W., and Vehrencamp, S.L. (1998). Principles of Animal Communication (Sinauer Associates Inc., Sunderland, MA).

Caswell, H. (1976). Community structure: a neutral model analysis. Ecol. Monogr. *46*, 327–354.

Chave, J. (2004). Neutral theory and community ecology. Ecol. Letts. *7*, 241–253

Chesson, P. (2000a). Mechanisms of maintenance of species diversity. Ann. Rev. Ecol. Sys. *31*, 343–366.

Chesson, P. (2000b). General theory of competitive coexistence in spatially-varying environments. Theor. Pop. Biol. *58*, 211–237.

Connell, J.H. (1978). Diversity in tropical rain forests and coral reefs. Science *199(4335)*, 1302–1310.

Dennett, D.C. (1995). Darwin's Dangerous Idea (Simon and Schuster, New York).

Etienne, R.S., and Rosindell, J. (2011). The spatial limitations of current neutral models of biodiversity. PLoS One *6*, e14717.

Fisher, R.A, Corbet, A.S., and Williams, C.B. (1943). The relation between the number of species and the number of individuals in a random sample of an animal population. J. Anim. Ecol. *12*, 42–58.

Forney, L.J., Zhou, X., and Brown, C.J. (2004). Molecular microbial ecology: land of the one-eyed king. Curr. Opin. Microbiol. *7*, 210–220.

Fridley, J.D., Peet, R.K., White, P.S., and Wentworth, T.R. (2005). Connecting fine- and broad-scale species–area relationships of Southeastern U.S. flora. Ecology *86*, 1172–1177.

Green, D.G., Klomp, N., Rimmington, G., and Sadedin, S. (2006). Complexity in Landscape Ecology (Springer, Dordrecht, the Netherlands).

Green, J.L., Hastings, A., Arzberger, P., Ayala, F.J., Cottingham, K.L., Cuddington, K., Davis, F., Dunne, J.A., Fortin, M.-J., Gerber, L., *et al.* (2005). Complexity in Ecology and Conservation: Mathematical, Statistical, and Computational Challenges. BioScience *55*, 501–510.

Grimm, V., and Wissel, C. (1997). Babel, or the ecological stability discussions: an inventory and analysis of terminology and a guide for avoiding confusion. Oecologia *109*, 323–334.

Harte, J., Zillio, T., Conlisk, E., and Smith, A.B. (2008). Maximum entropy and the state variable approach to macroecology. Ecology *89*, 2700–2711.

Harte, J., Smith, A.B., and Storch, D. (2009). Biodiversity scales from plots to biomes with a universal species–area curve. Ecol. Letts. *12*, 789–797.

Hubbell, S.P. (2001). The Unified Neutral Theory of Biodiversity and Biogeography. Monographs in Population Biology No. 32. (Princeton University Press, Princton, NJ).

Hughes, A.L. (2008). Near neutrality: leading edge of the neutral theory of molecular evolution. Ann. N.Y. Acad. Sci. USA. *1133*, 162–179.

Huston, M. (1979). A general hypothesis of species diversity. Am. Nat. *113*, 81–101.

Kimura, M. (1986). DNA and the neutral theory. Phil. Trans. R. Soc. Lond. B. *312*, 343–354.

Kimura, M. (1991). Recent development of the neutral theory viewed from the Wrightian tradition of theoretical population genetics. Proc. Natl. Acad. Sci. U.S.A. *88*, 5969–5973.

Kimura, M., and Crow, J.F. (1964). The number of alleles that can be maintained in a finite population. Genetics *49*, 725–738.

Leigh, E.G., Jr, Leigh, E.G., Wright, S.J., Herre, E.A., and Putz, F.E. (1993). The decline of tree diversity on newly isolated tropical islands: a test of a null hypothesis and some implications. Evol. Ecol. *7*, 76–102.

Ma, Z.S. (1991). Weibull distribution: as a spatial distribution model of insect population and its ecological interpretations. J. Biomathematics *6*, 34–45.

Ma, Z.S. (2011). Ecological 'theatre' for evolutionary computing 'play': some insights from population ecology and evolutionary ecology. I. J. Bio-Inspired Computing. *3*, 31–55.

MacArthur, R.H, and Wilson, E.O. (1967). The Theory of Island Biogeography (Princeton University Press, Princeton, NJ).

McGill, B.J. (2010). Towards a unification of unified theories of biodiversity. Ecol. Lett. *13*, 627–642.

McGill, B.J., and Brown, J.S. (2007). Evolutionary game theory and adaptive dynamics of continuous traits. Annu. Rev. Ecol. Evol. Syst. *38*, 403–435.

McGlinn, D.J., and Palmer, M.W. (2009). Modeling the sampling effect in the species–time–area relationship. Ecology *90*, 836–846.

Magurran, AE. (2004). Measuring Biological Diversity (Blackwell Publishing, Oxford, UK).

May, R.M. (1973). Stability and Complexity in Model Ecosystems. (Princeton University Press, Princeton, NJ).

Mueller, L.D., and Joshi, A. (2000). Stability in Model Populations. (Princeton University Press, Princeton, NJ).

Neutel, A.M., Heesterbeek, J.A., van de Koppel, J., Hoenderboom, G., Vos, A., Kaldeway, C., Berendse, F., and de Ruiber, P.C. (2007). Reconciling complexity with stability in naturally assembling food webs. Nature *449*, 599–602.

Preston, F.W. (1948). The commonness and rarity of species. Ecology *29*, 254–283.

Preston, F.W. (1962). The canonical distribution of commonness and rarity. Ecology *43*, 185–215.

Putman, R.J. (1994). Community Ecology (Chapman and Hall, New York).

Ravel, J., Gajer, P., Abdo, Z., Schneider, G.M., Koening, S.S., McCulle, S.L., Karlebach, S., Gorle, R., Russell, J., Tacket, C.O., *et al.* (2010). Vaginal microbiome of reproductive-age women. Proc Nat. Acad. Sci. *108(Suppl. 1)*, 4680–4687.

Robinson, C.J., Bohannan, B.J.M., and Young, V.B. (2010). From structure to function: the ecology of host-associated microbial communities. Microbiol. Mol. Biol. Rev. 74, 453–476.

Rosenzweig, M.L. (1998). Preston's ergodic conjecture: the accumulation of species in space and time. In Biodiversity Dynamics: Turnover of Populations, Taxa, and Communities, McKinney, M.L., and Drake, J.A., eds (Columbia University Press, New York), pp. 311–348.

Rosindell, J., and Cornell, S.J. (2009). Species-area curves, neutral models, and long-distance dispersal. Ecology *90*, 1743–1750.

Roxburgh, S.H., Shea, K., and Wilson, J.B. (2004). The intermediate disturbance hypothesis: Patch dynamics and mechanisms of species coexistence. Ecology *85*, 359–371.

Russell, G.J. (1998). Turnover dynamics across ecological and geological scales. In Biodiversity Dynamics: Turnover of Populations, Taxa, and Communities, McKinney, M.L., Drake, J.A., eds (Columbia University Press, New York), pp. 377–404.

Scheiner, S.M. (2003). Six types of species–area curves. Glob. Ecol. Biogeogr. *12*, 441–447.

Sousa, W.P. (1984). The role of disturbance in natural communities. Ann. Rev. Ecol. Sys. *15*, 353–391.

Southwood, T.R.E., and Henderson, P.A. (2000). Ecological Methods, 3rd Edition. (Wiley-Blackwell, UK).

Taylor, L.R., Kempton, R.A., and Woiwod, I.P. (1976). Diversity statistics and the log-series model. J. Anim. Ecol. *45*, 255–272.

Ugland, K.I., and Gray, J.S., and Ellingsen, K.E. (2003). The species-accumulation curve and estimation of species richness. J. Anim. Ecol. *72*, 888–897.

Watterson, G.A. (1974). Models for the logarithmic species abundance distribution. Theor. Popul. Biol. *6*, 217–250

White, E.P. (2004). Two-phase species–time relationships in North American land birds. Ecol. Lett. *7*, 329–336.

White, E.P., Ernest, S.K.M., Adler, P.B., Hurlbert, A.H., and Lyons, S.K. (2010). Integrating spatial and temporal approaches to understanding species richness. Phil. Trans. Royal Soc B: Biol. Sci. *365*, 3633–3643.

Woodcock, S., Christopher, J.G., Bell, T., Lunn, M., Curtis, T.P., Head, I.M., and Sloan, W.T. (2007). Neutral assembly of bacterial communities. FEMS Microbiol. Ecol. *62*, 171–180.

Wu, J., and Vankat, J.L. (1995). Island biogeography: theory and applications. In Encyclopedia of Environmental Biology. Vol. 2., Nierenberg, W.A., ed. (Academic Press, San Diego, CA), pp. 371–379.

Zengler, K. (2009). Central role of the cell in microbial ecology. Microbiol. Mol. Biol. Rev. *73*, 712–729.

Species–Time Relationships for Bacteria

5

Anna Oliver, Andrew K. Lilley and
Christopher J. van der Gast

Abstract

The identification of spatial patterns and their relationships to ecological events is an important specialization within ecology which is now branching into the microbial world. In spatial ecology, the detection of patterns at a given spatial scale can be used to explain ecological mechanisms and processes. Furthermore, through the application of spatial statistical analyses, factors leading to ecological events can be determined and verified. One of the most commonly studied aspects of spatial ecology, recently applied in microbial ecology, is the species–area relationship (SAR). The temporal analogue of the SAR, the species–time relationship (STR), on the other hand has received far less attention, even in the science of general ecology. Like SARs, the STRs are influenced by a variety of factors including dispersal, abiotic and biotic interactions, and species–species interactions. The application of these ecological conceptual tools to microbial ecology is a rapidly developing field. This chapter proposes that the STRs are a powerful and appropriate tool for studies of microbial diversity and that they make a contribution to understanding ecological communities. From a fundamental perspective, we focus on how microbial STRs compare with those for animals and plant communities, and how they are improving our understanding of community assembly and dynamics. As we believe a key future importance of studying STRs will be for applied benefit, we also discuss how microbial STRs have been used to distinguish between anthropogenic perturbations and underlying natural dynamics and have provided ecological insights for clinical benefit in bacterial infections.

Introduction

It is well known that microbial ecology is both driven and limited by the increasing plethora of techniques used to assess microorganisms and their communities. In many cases this has led to an almost unhealthy obsession for using the latest methodologies, typically at the expense of the research questions being asked. In comparison to animal or plant ecology, microbial ecology has been accumulating situation-bound statements of limited predictive ability which has resulted in a 'horizontal gain of information', rendering little insight to either researchers or practitioners (de Lorenzo, 2002; Prosser *et al.*, 2007). It has been previously argued that new technologies will increasingly lead us down 'blind non-generalist and expensive alleyways' if studies are not directed and driven by ecological theory (Prosser *et al.*, 2007). Given the central and global importance of microorganisms in natural and engineered ecosystems, progress requires the acceptance, development, and application of ecological theory and principles. However, the application of theory is still in its infancy in microbial ecology (Prosser *et al.*, 2007). The potential of exploiting theories, models and principles from general ecology, coupled with ever improving molecular methodologies, could well provide invaluable insights into how microbial communities organize and change in space and time. In time, this increased knowledge of microbial community ecology will help us better understand and predict changes in the natural environment, allow improved manipulation of agricultural and industrial processes and give improved protection of human health (van der Gast, 2008).

One of the fundamental objectives of ecology is to understand how biodiversity is accumulated and maintained across time and space. Patterns of species diversity provide important insights into the underlying mechanisms that regulate biodiversity, and are central to the development of ecological models and theories, such as the theory of island biogeography (MacArthur and Wilson, 1967) and the unified neutral theory of biodiversity and biogeography (Hubbell, 2001). One such pattern is the species–area relationship (SAR) which has provided the foundation for much theoretical ecology, and has been used to prioritize conservation efforts particularly in the SLOSS (a single large or several small reserves) debate (Gilpin and Diamond, 1980; Higgs and Usher, 1980; Simberloff and Abele, 1982).

Although spatial scaling of animal and plant species diversity has been well documented over the last century (e.g. Arrhenius, 1921; Preston, 1960; Connor and McCoy, 1979; Rosenzweig, 1995), it is only within the last decade that SARs have started to be addressed at the microbial level: algae (Smith *et al.*, 2005), bacteria (Horner-Devine *et al.*, 2004; Bell *et al.*, 2005; van der Gast *et al.*, 2005, 2006; Bell, 2010), fungi (Green *et al.*, 2004; van der Gast *et al.*, 2011a) and protozoa (Finlay, 2002). These studies were of fundamental importance as they provided evidence that different groups of microbial taxa had biogeographic structure in space thus refuting the long held tenet of microbial cosmopolitanism '*everything is everywhere*, but, *the environment selects*' (Baas Becking, 1934). Now firmly established, we believe the future importance of microbial SARs will lie in applied applications. For example, a recent applied study of spatial scaling of microbial diversity has demonstrated that arbuscular mycorrhizal fungi are affected by farming practice and local soil conditions at field and regional scales which has important implications for agriculture (van der Gast *et al.*, 2011a).

In contrast, the manner in which species richness changes with time has received less attention than spatial turnover (Rosenzweig, 1995). Temporal turnover can be defined as 'the number of species eliminated and replaced per unit time' and is the concept that is central to the theory of island biogeography (MacArthur and Wilson, 1967; Magurran, 2004). Like turnover in space it can be measured in a variety of ways as described by Magurran (2004). Here we will primarily focus our attention on the species–time relationship (STR) as a method to observe and contrast temporal turnover at the microbial level. Specifically, we will highlight how microbial STRs: (1) compare with those for animals and plant communities; and (2) are improving our understanding of community assembly and dynamics. As we believe a key future importance of studying STRs will be for applied benefit, we will also discuss how microbial STRs: (3) have been used to distinguish between anthropogenic perturbations and underlying natural dynamics; and (4) have provided ecological insights for clinical benefit in bacterial infections.

The species–time relationship

White (2004) stated that STRs are ecological patterns that describe the increase in observed taxa richness with length of time censored, and that studying STRs is important for understanding the ecological processes underlying temporal turnover and richness. Originally proposed by Preston (1960), the species–time relationship describes how the observed species richness of a community in a habitat of fixed size increases with the length of time over which the community is monitored (Preston, 1960; White, 2004; White *et al.*, 2006). The relationship between species richness and area is well described by the power law $S = cA^z$, where S is the number of observed species in area A, c is an empirically derived taxon- and location-specific constant and z is the slope of the line or spatial scaling exponent. Increasing values of z can be taken as greater rates of turnover or accumulation with area. The species-area power law can be modified to describe the relationship between species richness and time, T (Preston, 1960). For clarity, the scaling exponent z is changed to w, so the STR power law equation becomes $S = cT^w$ (Adler and Lauenroth, 2003; van der Gast *et al.*, 2008).

Rosenzweig (1995) advocated that the Gleason (semi-log) plot should be used to visualize the STR as he attempted to justify that the STR plots appeared to be a better fit by adjusting the abscissa (T) from an arithmetic scale to a logarithmic scale, where the STR model takes the form of $S = c + w \log T$ (Gleason, 1922). However, he conceded that the Arrhenius (log–log) plot worked equally well for visualizing/modelling a STR, where both the abscissa (T) and the ordinate (S) are plotted on logarithmic scales, where the STR model is of the form $\log S = \log c + w \log T$ or in the power law equation form $S = cT^w$ (Arrhenius, 1921). In agreement with Rosenzweig's concession, when both the semi-log and log–log versions of the STR were plotted using data from river water bacterial communities (Barnes *et al.*, 2010), there was no discernible difference between the two types of plot, in terms of the fit of the model to the data (Fig. 5.1). However, Preston (1960) proposed that the STR should mimic the SAR, following a straight line in log-log space, and would have similar scaling exponents to those of its spatial analogue as demonstrated for the log–log STR plots in Fig. 5.1. Therefore, we use the power law functional form for the STR, as this is how published studies of microbial taxa–area relationships have been plotted (van der Gast *et al.*, 2008; Barnes *et al.*, 2010).

Bacterial 'species' and sampling issues

A key obstacle microbial ecologists face in applying ecological theories and models to actual microbial communities is the shear extent of diversity in some systems and the difficulties of defining and counting, for example, bacterial species (van der Gast *et al.*, 2005). Although the unit of diversity in ecological models such as those used in the theory of island biogeography is frequently the species, it may be applied to higher or lower taxonomic levels. In practice there is not a secure bacterial species definition which is applicable to sampling complex microbial ecological communities. However, 16S rRNA sequencing has provided a valuable standard in phylogeny, which is applicable with available molecular methods. In many of the studies that will be discussed later in this chapter, a standard 16S rRNA gene sequence was polymerase chain reaction (PCR) amplified, to generate fragments for separation as distinct bands in denaturing gradient gel electrophoresis (DGGE), and was used to define operational taxonomic units (OTUs). The number of bands present in a sample is used to infer species richness (Muyzer *et al.*, 1993; Loisel *et al.*, 2006). The observed values

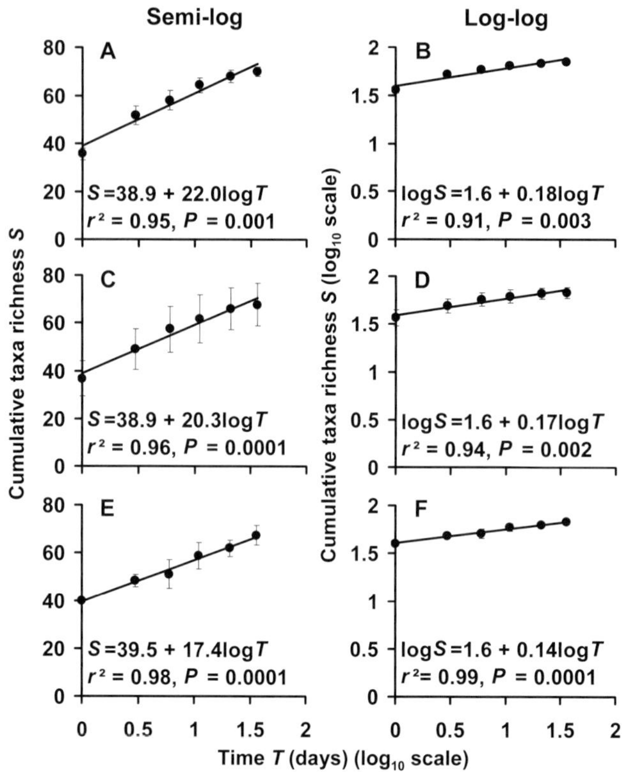

Figure 5.1 Species–time relationships (STR) for river water bacterial communities under different treatments visualized using semi-log (Gleason) and log–log (Arrhenius) plots. Treatments are: (A and B) river water (RW) plus iron nano-particles; (C and D) RW plus iron micro-particles; and (E and F) RW only. Given are the STR equations in semi-log and log–log form, coefficients of determination (r^2) and significance (*P*). Error bars represent the standard deviation of the mean ($n = 3$ in all cases). [Redrawn with permission from Barnes *et al*. 2010 (Copyright 2010, Elsevier).]

of the STR parameter log*c* (the intercept) will change with the level of OTU chosen. The different definitions of OTUs do lead to different richness values. However, these will result in stable values of *w* (scaling exponent), if different OTU definitions result in common proportional changes in taxa richness among the samples. The limitation of the traditional 'species' definition can thus be sidestepped, by defining species in terms of molecular identities. For the studies to be discussed, reproducible differences in the sequence of the 16S rRNA gene were used to 'type' bacteria, assign stable identities and assay diversity without resolving the taxonomic character of the types (Curtis *et al.*, 2002).

Interestingly, 'macrobial' and microbial ecologists share a similar problem, namely the ability to detect rarer species when sampling a given habitat (Magurran, 2004; Woodcock *et al.*, 2006). Although an anathema to most microbial ecologists, in a previous paper we drew upon the analogy Preston used 50 years previously to help define an ecological community (van der Gast *et al.*, 2005). Preston (1960) argued that rarer species tend to be, in ecological terms, non-essential to defining an ecological community and that, for example, a plant community could be defined by its more plentiful species with the rest being merely 'adventitious'. To emphasize the point further, he used the example of how petrographers define true granite; it must contain quartz, feldspar and mica (all essential minerals), while minor ingredients, such as sphene and zircon, are just 'accessory' minerals.

Woodcock *et al.* (2006) postulated that when high-throughput sequencing become routinely available (technically and financially), then a complete census of a sample may become available. In the interim, information on the richness, composition and structure of microbial communities is typically obtained using fingerprinting methods, such as terminal restriction fragment length polymorphism (T-RFLP) or DGGE, which exhibit method-dependent detection thresholds in absolute abundance below which rare taxa will not be detected; thus truncating the sample distribution (Woodcock *et al.*, 2006). Or rather we can only observe taxa that exist at a relative abundance above approximately 0.01 (Ofiteru *et al.*, 2010). Despite the known limitations of DGGE or T-RFLP, the patterns from such fingerprinting techniques allow valuable insights into the more dominant and therefore arguably the most ecologically significant proportion of a community by providing as Loisel and colleagues (2006) proposed, 'an image of the microbial ecosystem free from inventory limitation'.

Comparison of STRs across different groups of taxa

White *et al.* (2006) compared STRs for 984 community time-series of eukaryotic organisms (including: algae, zooplankton, invertebrates, insects, fish, birds, plants, mammals, and corals) from a range of aquatic and terrestrial ecosystems. They found a remarkable degree of regularity in the STR among ecosystems where temporal scaling exponents (w) ranged from 0.10–0.51 (Table 5.1), with a mean of $w = 0.29$. This is in accordance with the range of exponents from bacterial STR studies (w ~0.05–0.51) with a mean of 0.26 ($n = 72$) (Table 5.1). Although the slopes of the relationship are similar to that found in eukaryotic organism groups, it remains to be seen whether the same processes are responsible for structuring both bacterial and eukaryotic communities. Furthermore, it is notable that the exponents are similar despite different lengths in time series and that microbial, animal and plant communities have similar exponent values. A possible explanation for this may be a general scaling relationship with body size, generation times, dispersal, extinction (local) and turnover. It is important to mention, however, that there are a limited number of published STRs for bacteria and as increasing STRs for different taxonomic groups are published a more informative picture may well emerge.

Community assembly and dynamics

A fundamental objective of ecology is to understand how ecological communities form and are maintained across spatial and temporal scales. In broad terms, species that assemble to make up a community are determined by (1) dispersal constraints; (2) environmental constraints (abiotic context, habitat heterogeneity); and (3) internal dynamics (e.g. mutualism, parasitism, predation, and competition) (Begon *et al.*, 2006). As a result, ecologists search for rules of community assembly in order to understand and better predict this important process (Begon *et al.*, 2006). At the microbial level a number of studies, primarily based in engineered systems, have used species–time relationships to help test whether theoretical

Table 5.1 A comparison of species–time relationships across different taxonomic groups

Taxonomic group	STR slope (*w*)	No. of STRs	Time series length	Reference
Eukaryotes				
Invertebrates	0.23–0.40	98	10–23 years	White *et al.* (2006)
Vertebrates	0.10–0.51	637	11–27 years	White *et al.* (2006)
Land plants	0.32–0.46	164	14–35 years	White *et al.* (2006)
Algae	0.21–0.30	85	13–23 years	White *et al.* (2006)
Bacteria				
Activated sludge bacteria	0.26 (± 0.08)	12	39 days	Ayarza and Erijman (2011)
Activated sludge bacteria	0.35 (± 0.15)	5	154 days	van der Gast *et al.* (2008)
Cystic fibrosis bacteria	0.19 (± 0.08)	30	20 min	Rogers *et al.* (2010a)
Leaf bacteria	0.31	1	Circa 160 days	Redford and Fierer (2009)
River water bacteria	0.17 (± 0.02)	9	36 days	Barnes *et al.* (2010)
Sump tank bacteria	0.14 (± 0.10)	6	49 days	van der Gast (2008)
Tree hole bacteria	0.28 (± 0.12)	9	24 days	Ager *et al.* (2010)

predictions for community assembly and dynamics are applicable to bacterial communities. These include the effects of habitat size, selection pressure, and metacommunity size on local communities. Using such systems, as employed in the studies described below, it has become popular to test different aspects of ecological theory at the microbial level. Bacterial communities are readily identified or assembled and are considered analogous to island habitats in which experimental environmental factors (including temperature, nutrient availability and type, habitat size, etc.) can be conveniently manipulated and controlled (van der Gast *et al.*, 2006). Islands have proven well suited to provoking or testing theories in general ecology as they contain tractable communities with defined boundaries (Diamond and May, 1976). Indeed, they have played a key role in the development of ecological theories, from the initial proposition of evolution and biogeography with Darwin's finches on the Galápagos Islands (Darwin, 1859), through Mayr's demonstration of the role of geographic isolation in speciation (Mayr, 1963), to the theory of island biogeography pioneered by MacArthur and Wilson (1967).

Island size

Theoretical predictions for island biogeography state that smaller islands will have lower (less stable) diversity, whereas larger islands will have greater and more stable diversity over time (Fig. 5.2A). This has primarily been explained by larger islands having more refuge opportunities than smaller ones (MacArthur and Wilson, 1967; Curtis *et al.*, 2003). Data from two studies of three metalworking fluid sump tanks (small, medium, and large islands) sampled every 7 days for 8 weeks were used to test this prediction (van der Gast *et al.*, 2005; van der Gast, 2008). Plotting the STR for each of the large, medium

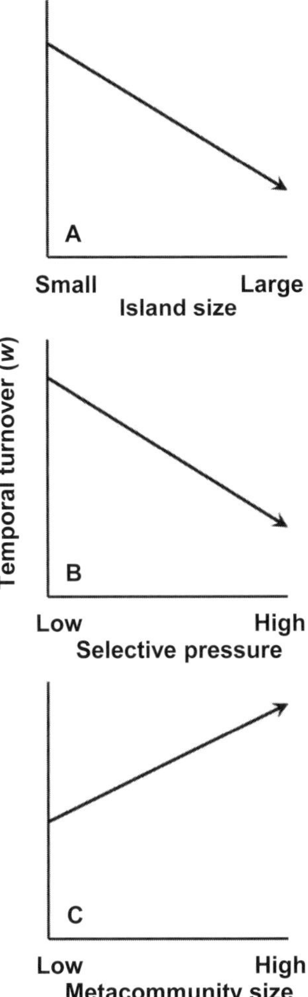

Figure 5.2 The predicted effects of (A) island size, (B) selective pressure, and (C) metacommunity size on temporal turnover in bacterial communities using the species–time relationship scaling exponent *w* as the measure of turnover.

and small sump tank 'islands', it was observed that the large sump tanks had low temporal scaling exponents ($w = 0.09$ and 0.02) suggesting that bacterial taxa turnover was low on these islands. The medium sized islands fell between the observed properties of large and small islands ($w = 0.11$ and 0.12). Whereas, small islands had much steeper STR scaling exponents ($w = 0.23$ and 0.29) indicating that small islands have less stable bacterial populations and subsequently turnover was much greater (van der Gast, 2008). In agreement with this, cluster analyses from both studies revealed that bacterial communities over time in the large sump tanks were most similar in composition. In contrast, small islands showed less similarity across time, and again the medium size islands were intermediate in similarity (van der Gast *et al.*, 2005).

Selection pressure

A central question in ecology is why so many species can coexist when there are so few different apparent niches in a given habitat. Ecologists have attempted to explain this by understanding how communities assemble. This in turn has led to two opposed opinions of deterministic or stochastic assembly. The deterministic view asserts that communities assemble in a defined manner as a result of the number of available niches and competition for those niches in a habitat (Diamond, 1975; McGill, 2003). Conversely, the stochastic view assumes that assembly is random, making neutral assumptions about individuals within a community, where neutrality is defined as ecological equivalence among all individuals of every species (Hubbell, 2001). Put simply, all individuals have the same chance of immigrating into the local community, and reproducing or going extinct once there.

In a previous study, van der Gast *et al.* (2008) sampled five bioreactors of fixed volume size along a gradient of increasing industrial wastewater concentrations (0%, 25%, 50%, 75% and 100%) and consequently decreasing municipal wastewater over a 154-day period. The rationale for the study was to ascertain how bacterial taxa scaled temporally under increasing selective pressure exerted by the wastewater blends in the bioreactor systems, and to what extent bacterial community dynamics were driven by either stochastic or deterministic considerations across the established selective pressure gradient. The authors found that as the industrial wastewater increased in concentration across the bioreactors, a gradual switch from stochastic (random) community assembly to more deterministic (niche) based considerations was observed. This was also reflected in differences in STR scaling exponents (w), where w decreased as selective pressure (industrial wastewater) increased [0% ($w = 0.512$), 25% (0.432), 50% (0.315), 75% (0.206) and 100% (0.162); van der Gast *et al.*, 2008]. These results would suggest that as selective pressure and therefore the level of determinism in community assembly increases then temporal turnover of bacterial taxa decreases as a result (Fig. 5.2B). What is indicated from this work is that a continuum of both deterministic and stochastic processes can exist by imposing a gradient of selective pressure upon the bacterial communities within the bioreactors. The wider implications are that purely neutral community assembly models should incorporate the influence of deterministic factors and vice versa as recently performed by Ofiteru *et al.* (2010), who combined niche and neutral effects in a model for ammonia-oxidizing bacteria involved in wastewater treatment.

Metacommunity size

Following on from the work of van der Gast and colleagues (2008), Ayarza and Erijman (2011) also indicated that the dynamics of activated sludge bacterial communities are determined by a balance between neutral and deterministic components however they used a different line of evidence to demonstrate this. By manipulating the size of the meta or source community used to form local bioreactor communities they showed, using the goodness of fit to a neutral community model modified for prokaryotes (Sloan *et al.*, 2006), that the higher the number of species in the reservoir from which the local community is drawn, the more important the stochastic component is in the formation of the activated sludge communities. Conversely, when the source community diversity was lower, the more important the deterministic component is in community assembly. To verify this, they also analysed the effect of source community size on the rate of temporal turnover. They found that turnover was greater in bioreactor communities ($w = 0.31 \pm 0.05$) formed from more diverse metacommunities when compared with local communities formed from less diverse source communities ($w = 0.16 \pm 0.02$) (Ayarza and Erijman, 2011). In short, they found that the larger the metacommunity size the greater the turnover and stochasticity in the dynamics of the local communities (Fig. 5.2C).

Effects of anthropogenic disturbances

One of the great scientific challenges that we currently face is how to mitigate threats to the environment and biodiversity. Potential threats arise from any anthropogenic disturbances including; climate change, fragmentation and loss of habitats, pollution events and many other factors leading to environmental change. Detecting and quantifying biodiversity is a fundamental prerequisite to understanding how ecosystems and their services are formed and how any disturbance may affect ecosystem health. Furthermore, before attempting mitigation it is important to address how to evaluate biological integrity and recovery following an anthropogenic induced stress (Ager *et al.*, 2010).

Ecologists have become increasingly aware that there is limited knowledge for temporal ecological communities and the changes they undergo (Magurran *et al.*, 2010). With an unprecedented concern about biodiversity and environmental change it is essential to monitor biodiversity, and to gauge changes in biodiversity through time. Specifically, as all communities experience temporal turnover, a major challenge is to distinguish change that can be attributed to external factors from underlying natural community dynamics (Magurran *et al.*, 2010). Encouragingly, this issue for animal and plant species is starting to be addressed through the use of existing long-term data sets from biodiversity research and monitoring networks, and has real potential for assessing change in ecological communities through time (Magurran *et al.*, 2010). Even so, incorporation of temporal turnover into environmental microbiology is still in its first phase (Ager *et al.*, 2010).

To address this at the microbial level, Ager *et al.* (2010) used water-filled tree holes as a model system for studying the effects of disturbances upon bacterial community STRs. Using pentachlorophenol (PCP) as a model pollutant, they assessed the effects of three different levels of treatment [strong perturbation (500 ppm PCP), intermediate (50 ppm), and no perturbation (control)] on bacterial community structure. This was contrasted with other community characteristics including richness, composition and temporal turnover, over time in replicated experimental tree hole microcosms. Between days 0 and 1 of the 24-day study, PCP was dosed into replicate tree hole microcosms at 500 or 50 ppm concentrations. In addition, unperturbed (control) microcosms were also sampled over the same study period to determine the impact of anthropogenic perturbation from the underlying natural community dynamics.

It is general accepted in ecology that a species abundance distribution for a given community of animals or plants following a perturbation will typically change in structure from one of relative evenness to one of increased dominance. Subsequently, such changes in evenness have been used as indicators of biological integrity and environmental assessment. The main purpose of Ager and colleague's (2010) study was primarily to investigate whether changes in bacterial community structure would reflect the effects of anthropogenic stress in a similar manner to larger organisms. Their findings demonstrated that the bacterial community structure observed reflected both impact and recovery from an anthropogenic disturbance indicating that this may well be a potentially useful metric for environmental assessment. These findings were also reflected in the temporal turnover of microcosms, as the mean temporal scaling exponents were observed to increase with the level of disturbance (Fig. 5.3).

Using a similar experimental design and approach, Barnes *et al.* (2010) assessed the impacts of iron nano-particles and iron micro-particles on river water bacterial communities. They found that, when compared with control river water microcosms, the addition of iron nano- or micro-particles did not influence bacterial community structure (Barnes *et al.*, 2010). Likewise, this was reflected when temporal turnover of bacterial taxa across the treatments was assessed using STRs (Fig. 5.1). An increase in turnover would be expected following a perturbation or environmental stress event. However, the mean temporal scaling exponents did not increase with disturbance, and the mean values of the slopes were not significantly different across treatments. Therefore, it was concluded that addition of iron nano- or micro-particles did not affect the river water bacterial communities.

Ecological insights for clinical benefit

Cystic fibrosis (CF) patients suffer from chronic bacterial lung infections that lead to death in the majority of cases. The need to maintain lung function in these patients means

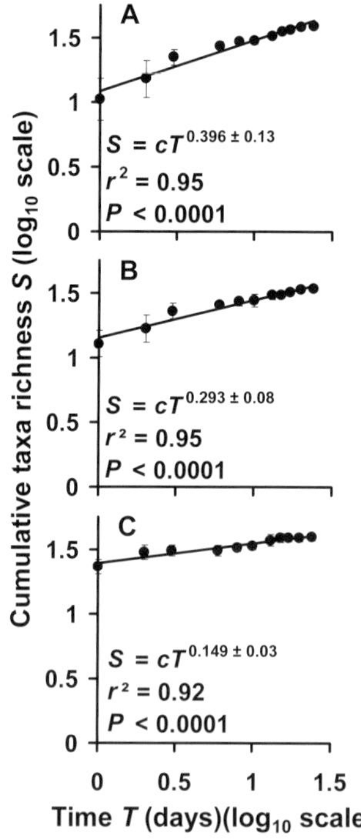

Figure 5.3 Species–time relationships for bacterial communities in water-filled tree hole microcosms under different levels of pentachlorophenol (PCP) perturbation. Treatments are (A) 500 pm PCP, (B) 50 ppm PCP, and (C) no perturbation (control). Given are the STR power law equation ($S = cT^w$), coefficients of determination (r^2), and significance (P). [Redrawn with permission from Ager *et al.* 2010 (Copyright 2010, John Wiley and Sons).]

characterizing these infections is vital (van der Gast *et al.*, 2011b). The adoption of culture-independent methodologies developed to address environmental microbiological questions have revealed the presence of much more complex bacterial communities than previously realized (Rogers *et al.*, 2005, 2009). Work on CF respiratory microbiology has led to the proposal of novel important concepts. Extending from observations of bacterial diversity in the CF lower airways, these infections have recently been proposed to be driven by communities of interacting organisms (Rogers *et al.*, 2010b). As such, the chronically colonized CF airways represent a previously unrealized complex and diverse ecosystem (Harrison, 2007). Therefore the advantages in structuring subsequent research within an ecological framework are clear, as ecological insights can provide real clinical benefit for such chronic pulmonary infections (van der Gast *et al.*, 2011b).

Selection of appropriate treatment for CF airway infections requires knowledge of what bacterial species are present in a sample. However, to avoid bias and for this to provide information for clinical use, it is essential that the material (sputum) sampled and analysed reflect the composition of secretions in the lower airways and that this material be collected in a reproducible manner. Recently, Rogers and colleagues (2010) analysed sputum samples provided serially at 5-min intervals over a 20-min period for each patient from a cohort of ten adult patients; with an aim to determine whether one sample, the standard diagnostic method, or a series of samples provided over a 20-min period would reflect the bacterial diversity and community composition within each patient. For all patients, species–time relationships reached, or were about to reach, an asymptote at the 20-min time point, suggesting that enough diversity had been sampled and that further bacterial species were unlikely to be detected through collection of further samples (Rogers *et al.*, 2010a). In addition, if only a single sample was analysed, the proportion of total species richness detected was calculated to be 58.7 ± 19.8% ($n = 30$) and therefore would underestimate diversity of bacterial infections within each patient from the cohort sampled.

Summary and outlook

Preston (1960) was the first to propose that SARs and STRs are similar in mathematical form, and Rosenweig (1998) united these two types of species richness relationships (SRRs) further, arguing that similar underlying mechanisms create both patterns. STRs together with other SRRs such as SARs provide powerful tools for the assay of community diversity (Scheiner *et al.*, 2011), and are set to play a key part in characterizing and comprehending microbial diversity. The critical question is to identify the mechanisms and processes by which the species count rises with time. Scheiner *et al.* (2011) stated that SRRs do not directly distinguish among mechanisms, as particular relationships can be the result of different combinations of causes. However, a better understanding of how any particular mechanism may contribute to the form or shape of SRRs allows important deductions to be made about ecological processes from patterns.

In a recent landmark paper, Scheiner and colleagues (2011) examined and discussed the underpinnings of species richness relationships in space and time, noting that a variety of ecological mechanisms influence the shapes of any particular SRR. Specifically, they proposed that those mechanisms can be gathered within five broad classes of (1) more individuals [by including more sampling units (a larger area or a longer time), more individuals are invariably sampled]; (2) environmental heterogeneity (as sampling increases to include more area or more time, more environmental variation will be encountered); (3) dispersal limitations (SRRs are affected by the movement and dispersal limits of individuals and species); (4) biotic interactions (similar to dispersal, biotic interactions can influence the degree of within- and between-species aggregation); and (5) multiple species pools (or metacommunities) (Scheiner *et al.*, 2011). Certainly, their approach could be employed as an opening framework to elucidate which mechanisms influence the form of microbial STRs or SARs and to some extent this has been touched upon here in the section on bacterial community assembly and dynamics in terms of habitat size, selective pressure, and source community size.

At the microbial level, a simple starting point is that as the number of bacteria sampled increases so does the number of species detected. However, this needs to be teased apart into additional species detected with larger sample size and those additional species which are a product of random immigration, seasonal or successional processes. Seasonal or successional community changes, where the microbial community modifies its abiotic and biotic environment as it progresses, are adaptive and likely to show changes which correlate with other ecological data.

Unlike plants and animals, bacterial communities can significantly benefit from increases in diversity through ecological time scale evolution. It is well established that bacteria are capable of restructuring their genetic structure to colonize new niches and take advantage of different nutrient sources. The literature is now legion with such data including the effects of genome evolution and horizontal gene transfer. For an interesting example see Smith *et al.* (2006), where *Pseudomonas aeruginosa* in CF lung bacterial communities go through substantial genetic adaptations, frequently generating a variety of strains with different traits (Smith *et al.*, 2006). As the molecular tools available to microbial ecology improve, it may well be possible to further examine the influence on the scaling exponent, w, by moving to the strain level and incorporating the real time accumulation of genetic diversity.

With their small body sizes, high densities, and potentially fast growth rates it is to be expected that microbial studies will have some distinct differences from most plant and animal studies. One key difference to be born in mind is that microbial species detected at subsequent time points may have been previously present. There is therefore some ambiguity about the location of the metacommunity from which the local community acquires its newly detected species. This issue is not absent from plant and animal studies, where immigrants may well be relatively rare and therefore remain undetected for some time. Their eventual detection may be, with time, as 'rare species' or result from their subsequent proliferation. This will occur in plant surveys when the seed bank retains species which are not continually present. The species-time power law relationships observed here are probably a composite of the temporal processes by which bacteria come to prominence as first inward immigrants and then proliferation *in situ*. These may then be followed by successive losses or declines followed by new proliferations.

The initial research detailed here on STRs should, as previously stated by White *et al.* (2006), highlight the dynamic nature of ecological communities. It should also emphasize a need to incorporate temporal scaling into both basic and applied research on species richness patterns (White *et al.*, 2006). As is the usual conclusion from most studies, more research is needed for temporal scaling of bacterial diversity to better understand the underlying mechanisms that form these patterns. In the interim, we believe the STR to be a powerful method for observing temporal turnover within bacterial communities. In addition, we believe the main applications will be in better understanding community assembly and as a tool to measure and assess anthropogenic impacts on microbial communities by distinguishing such change from underlying community dynamics.

Acknowledgements

Anna Oliver and Christopher J. van der Gast are supported by the UK Natural Environmental Research Council (NERC). Andrew K. Lilley is supported by the Wellcome Trust.

References

Adler, P.B., and Lauenroth, W.K. (2003). The power of time: Spatiotemporal scaling of species diversity. Ecol. Lett. *6*, 749–756.

Ager, D., Evans, S., Li, H., and van der Gast, C.J. (2010). Anthropogenic disturbance affects the structure of bacterial communities. Environ. Microbiol. *12*, 670–678.

Arrhenius, O. (1921). Species and area. J. Ecol. *9*, 95–99.

Ayarza, J.M., and Erijman, L. (2011). Balance of neutral and deterministic components in the dynamics of sludge floc assembly. Microb. Ecol. *61*, 486–495.

Baas Becking, L.G.M. (1934). Geobiologie of Inleiding tot de Milieukunde (W.P. Van Stockum & Zoon, The Hague) (in Dutch).

Barnes, R.J., van der Gast, C.J., Riba, O., Prosser, J.I., Dobson, P.J., and Thompson, I.P. (2010). The impact of zero-valent iron nanoparticles on a river water bacterial community. J. Hazard. Mater. *184*, 73–80.

Begon, M., Townsend, C.R., and Harper, J.L. (2006). Ecology: From Individuals to Ecosysytems, 4th edn (Blackwell Publishing, Oxford, UK).

Bell, T. (2010). Experimental tests of the bacterial distance–decay relationship. ISME J. *4*, *1357–1365*.

Bell, T., Ager, D., Song, J.-I., Newman, J.A., Thompson, I.P., Lilley, A.K., and van der Gast, C.J. (2005). Larger islands house diverse bacterial taxa. Science *308*, 1884.

Connor, E.F., and McCoy, E.D. (1979). The statistics and biology of the species–area relationship. Am. Nat. *113*, 791–833.

Curtis, T.P., Sloan, W.T., and Scannell, J.W. (2002). Estimating prokaryotic diversity and its limits. Proc. Natl. Acad. Sci. U.S.A. *99*, 10494–10499.

Curtis, T.P., Head, I.M., and Graham, D.W. (2003). Theoretical ecology for engineering biology. Environ. Sci. Technol. *37*, 64A–70A.

Darwin, C. (1859). On the Origin of Species by Means of Natural Selection, or the Preservation of Favoured Races in the Struggle for Life (John Murray, London).

Diamond, J.M. (1975). Assembly of species communities. In Ecology and Evolution of Communities, Cody, M.L., and Diamond, J.M., eds (Harvard University Press, Cambridge, MA), pp. 342–444.

Diamond, J.M., and May, R.M. (1976). Island biogeography and the design of natural reserves. In Theoretical Ecology: Principles and Applications, R.M. May, ed. (Oxford, Blackwell Scientific), pp. 163–186.

Finlay, B.J. (2002). Global dispersal of free-living microbial eukaryote species. Science *296*, 1061–1063.

van der Gast, C.J. (2008). Islands shaping thought in microbial ecology. Adv. Appl. Microbiol. *64*, 167–182.

van der Gast, C.J., Lilley, A.K., Ager, D., and Thompson, I.P. (2005). Island size and bacterial diversity in an archipelago of engineering machines. Environ. Microbiol. *7*, 1220–1226.

van der Gast, C.J., Jefferson, B., Reid, E., Robinson, T., Bailey, M.J., Judd, S.J., and Thompson, I.P. (2006). Bacterial diversity is determined by volume in membrane bioreactors. Environ. Microbiol. *8*, 1048–1055.

van der Gast, C.J., Ager, D., and Lilley, A.K. (2008). Temporal scaling of bacterial taxa is influenced by both stochastic and deterministic ecological factors. Environ. Microbiol. *10*, 1411–1418.

van der Gast, C.J., Gosling, P., Tiwari, B., and Bending, G.D. (2011a). Spatial scaling of arbuscular mycorrhizal fungal diversity is affected by farming practice. Environ. Microbiol. *13*, 241–249.

van der Gast, C.J., Walker, A.W., Stressmann, F.A., Rogers, G.B., Scott, P., Carroll, M.P., Parkhill, J., and Bruce, K.D. (2011b). Partitioning core and satellite taxa from within cystic fibrosis lung bacterial communities. ISME J. *5*, 780–791.

Gilpin, M.E., and Diamond, J.M. (1980). Subdivision of nature reserves and the maintenance of species diversity. Nature *285*, 567–568.

Gleason, H.A. (1922). On the relation between species and area. Ecology *71*, 213–225.
Green, J.L., Holmes, A.J., Westoby, M., Oliver, I., Briscoe, D., Dangerfield, M., Gillings, M., and Beattie, A.J. (2004). Spatial scaling of microbial eukaryote diversity. Nature *432*, 747–750.
Harrison, F. (2007). Microbial ecology of the cystic fibrosis lung. Microbiology *153*, 917–923.
Higgs, A.J., and Usher, M.B. (1980). Should nature reserves be large or small? Nature *285*, 568–569.
Horner-Devine, M.C., Lage, M., Hughes, J.B., and Bohannan, B.J.M. (2004). A taxa–area relationship for bacteria. Nature *432*, 750–753.
Hubbell, S.P. (2001). The Unified Neutral Theory of Biodiversity and Biogeography (Princeton University Press, Princeton, NJ).
Loisel, P., Harmand, J., Zemb, O., Latrille, E., Lobry, C., Delgenes, J.-P., and Godon, J.-J. (2006). Denaturing gradient electrophoresis (DGE) and single-strand conformation polymorphism (SSCP) molecular fingerprints revisited by simulation and used as a tool to measure microbial diversity. Environ. Microbiol. *8*, 720–731.
de Lorenzo, V. (2002). Towards the end of experimental (micro)biology? Environ. Microbiol. *4*, 6–9.
MacArthur, R.H., and Wilson, E.O. (1967). The Theory of Island Biogeography (Princeton University Press, Princeton, NJ).
McGill, B.J. (2003). A test of the unified neutral theory of biodiversity. Nature *422*, 881–885.
Magurran, A.E. (2004). Measuring Biological Diversity (Blackwell Publishing, Oxford, UK).
Magurran, A.E., Baillie, S.R., Buckland, S.T., Dick, J.M., Elston, D.A., Scott, E.M., Smith, R.I., Somerfield, P.J., and Watt, A.D. (2010). Long-term datasets in biodiversity research and monitoring: assessing change in ecological communities through time. Trends. Ecol. Evol. *25*, 574–582.
Mayr, E. (1963). Animal Species and Evolution (Harvard University Press, Cambridge, MA).
Muyzer, G., Dewaal, E.C., and Uitterlinden, A.G. (1993). Profiling of complex microbial-populations by denaturing gradient gel electrophoresis analysis of polymerase chain reaction-amplified genes coding for 16S ribosomal RNA. Appl. Environ. Microbiol. *59*, 695–700.
Ofiteru, I.D., Lunn, M., Curtis, T.P., Wells, G.F., Criddle, C.S., Francis, C.A., and Sloan, W.T. (2010). Combined niche and neutral effects in a microbial wastewater treatment community. Proc. Natl. Acad. Sci. U.S.A. *107*, 15345–15350.
Preston, F.W. (1960). Time and space and the variation of species. Ecology *41*, 611–627.
Prosser, J.I., Bohannan, B.J.M., Curtis, T.P., Ellis, R.J., Firestone, M.K., Freckleton, R.P., Green, J.L., Green, L.E., Killham, K., Lennon, J.J., *et al.* (2007). The role of ecological theory in microbial ecology. Nat. Rev. Microbiol. *5*, 384–392.
Redford, A.J., and Fierer, N. (2009). Bacterial succession on the leaf surface: A novel system for studying successional dynamics. Microb. Ecol. *58*, 189–198.
Rogers, G.B., Carroll, M.P., Serisier, D.J., Hockey, P.M., Kehagia, V., Jones, G.R., and Bruce, K.D. (2005). Bacterial activity in cystic fibrosis lung infections. Respir. Res. *6*, 49–60.
Rogers, G.B., Daniels, T.W.V., Tuck, A., Caroll, M.P., Connett, G.J., David, G.J.P., and Bruce, K.D. (2009). Studying bacteria in respiratory specimens by using conventional and molecular microbiological approaches. BMC Pulm. Med. *9*, 14–24.
Rogers, G.B., Skelton, S., Serisier, D.J., van der Gast, C.J., and Bruce, K.D. (2010a). Determining cystic fibrosis-affected lung microbiology: Comparison of spontaneous and serially induced sputum samples by use of terminal restriction fragment length polymorphism profiling. J. Clin. Microbiol. *48*, 78–86.
Rogers, G.B., Stressmann, A., Walker, A.W., Caroll, M.P., and Bruce, K.D. (2010b). Lung infections in cystic fibrosis: deriving clinical insight from microbial complexity. Expert Rev. Mol. Diagn. *10*, 187–196.
Rosenzweig, M.L. (1995). Species Diversity in Space and Time (Cambridge University Press, Cambridge, UK).
Rosenzweig, M.L. (1998). Preston's ergodic conjecture: The accumulation of species in space and time. In Biodiversity Dynamics: Turnover of Populations, Taxa, and Communities, McKinney, M.L., and Drake, J.A., eds (Columbia University Press, New York), pp. 311–348.
Scheiner, S.M., Chiarucci, A., Fox, G.A., Helmus, M.R., McGlinn, D.J., and Willig, M.R. (2011). The underpinnings of the relationship of species richness with space and time. Ecol. Monogr. *81*, 195–213.
Simberloff, D., and Abele, L.G. (1982). Refuge design and island biogeographic theory: effects of fragmentation. Am. Nat. *120*, 41–50.
Sloan, W.T., Lunn, M., Woodcock, S., Head, I.M., Nee, S., and Curtis, T.P. (2006). Quantifying the roles of immigration and chance in shaping prokaryote community structure. Environ. Microbiol. *8*, 732–740.
Smith, E.E., Buckley, D.G., Wu, Z.N., Saenphimmachak, C., Hoffman, L.R., D'Argenio, D.A., Miller, S.I., Ramsey, B.W., Speert, D.P., Moskowitz, S.M., *et al.* (2006). Genetic adaptation by *Pseudomonas aeruginosa* to the airways of cystic fibrosis patients. Proc. Natl. Acad. Sci. U.S.A. *103*, 8487–8492.

Smith, V.H., Foster, B.L., Grover, J.P., Holt, R.D., Leibold, M.A., and deNoyelles, F. (2005). Phytoplankton species richness scales consistently from laboratory microcosms to the world's oceans. Proc. Natl. Acad. Sci. U.S.A. *102*, 4393–4396.

White, E.P. (2004). Two-phase species–time relationships in North American land birds. Ecol. Lett. *7*, 329–336.

White, E.P., Adler, P.B., Lauenroth, W.K., Gill, R.A., Greenberg, D., Kaufman, D.M., Rassweiler, A., Rusak, J.A., Smith, M.D., Steinbeck, J.R., *et al.* (2006). A comparison of the species–time relationship across ecosystems and taxonomic groups. Oikos *112*, 185–195.

Woodcock, S., Curtis, T.P., Head, I.M., Lunn, M., and Sloan, W.T. (2006). Taxa–area relationships for microbes: the unsampled and unseen. Ecol. Lett. *9*, 805–812.

Microbial Biogeography: Is Everything Small Everywhere?

Diego Fontaneto and Joaquín Hortal

Abstract

The distribution of microscopic organisms (that is, those smaller than 2 mm) has been historically considered non-relevant for biogeography, because of the idea that owing to their small size, dispersal abilities, resting stages and quick reproductive rates, the presence of microscopic organisms in any place was not limited by geographical barriers and distances. Recent studies challenge this idea, and provide theoretical and empirical evidence in support of the existence of spatial patterns at different scales, and of biogeographical processes affecting many groups of microscopic organisms. Here we review the current state of the art for microbial biogeography, summarizing sources of problems and misconceptions, but also their solutions advancing the general understanding of biogeography, and conclude suggesting new avenues for future research.

Introduction

Microbial biogeography, the study of the distribution of microscopic organisms (smaller than 2 mm), is a relatively young discipline. Although the first attempts to describe microbial distribution date from the first decades of the Twentieth Century, the topic has not received much attention until the 1990s. Only recently, microbial biogeography has gained renewed vigour because of the resurgence of the 'Everything-is-Everywhere' hypothesis (Finlay, 2002; Fenchel and Finlay 2003; Finlay and Esteban, 2007; for historical reviews of this idea see O'Malley, 2008; Williams, 2011). We will not deal with the discussion on whether it is a true hypothesis in strict terms, as 'regardless of detail or explanation, the phrase 'everything is everywhere' is clearly false, and in any case, one would need to know in what sense the words 'everything' and 'everywhere' are intended. As soon as something is not found everywhere then everything (obviously) is not everywhere. This much could be easily agreed upon' (quotation from Williams, 2011). Notwithstanding ambiguities and semantic problems, this idea produced interesting theoretical and empirical studies, which we will summarize in the next sections, dealing with what we know about the effect of size, taxonomy and space in microbial biogeography; then, we describe the empirical evidence on different group of organisms and conclude with potential consequences and challenges. Knowing if and how microscopic organisms attain global distributions is a far reaching topic, going further than biogeography itself. For example, is speciation happening also in the absence of geographical barriers that would isolate populations? Here we introduce and discuss this and other key topics of current research in microbial biogeography, summarizing the recent review of Fontaneto (2011).

Size

Size is one of the major drivers of most biological properties. The size of a living organism largely determines its shape, how it interacts with the environment and most (if not all) biological functions or its life history (Calder, 1996; Bonner, 2006). Importantly, the differences in body size result in profound biological differences, because the relationship between the volume and surface of organisms is non-linear. Small increments in body surface result in large increments in volume; as a consequence, larger organisms have to regulate their temperature, osmosis, physiology, metabolism in a radically different way than small ones (Schmidt-Nielsen, 1984; Peters, 1986). Body size also plays a major role in determining the ecology of each species or individual, from their behaviour to the communities they interact with or the ecosystems where they live in, and even their extinction risk (Colinvaux, 1978; Cardillo and Bromhan, 2001; Hildrew *et al.*, 2007). As a consequence, current research in many areas attempts to develop empirical and/or theoretical models relating biological patterns and processes with body size. This has resulted in the development of a large body of theory on the relationship between size and metabolism, based on universal scaling laws (Brown and West, 2000; Brown *et al.*, 2004). Although to date such theory has failed to describe natural patterns when put to the test (e.g. Muller-Landau *et al.*, 2006; Hawkins *et al.*, 2007; Martínez del Rio, 2008), current knowledge on many of the biological constraints imposed by body size is fairly good.

A notable exception to this general trend is the biogeographical implication of body size. Geographical variations in the body size of large organisms are relatively well known. As early as the mid Nineteenth Century, Bergmann (1847) proposed that populations of endothermic animals located at different latitudes differ in their body mass, increasing with latitude and colder climate (the so-called Bergmann's rule; see Meiri and Dayan, 2003). This formulation has been extended to the species level, as well as to ectotherms (e.g. Olalla-Tárraga *et al.*, 2006); in short, for many animal groups within closely related species the larger in size will be found at higher latitudes. The exact mechanisms behind this common (but not universal) trend are still elusive, for the origin of such latitudinal gradient may be related to many factors and to past climate changes, rather than just to current climate (Diniz-Filho *et al.*, 2009). This rule, and many of the biogeographical patterns already described for large organisms do not scale down to microscopic organisms. Rather than a gradient in the patterns of diversity from large (macrobes) to small organisms (microbes), the general idea is that there may be an abrupt distinction between their biogeographical patterns.

The differences in the biogeographical patterns shown by microbes and macrobes have led to the hypothesis of a strong distinction in patterns and processes between macroscopic organisms with biogeography and microscopic organisms without biogeography (Finlay, 2002; Fenchel and Finlay, 2004). The threshold distinguishing these two size groups has been empirically defined to fall somewhere between 1 and 10 mm, although there is a general agreement that 2 mm would be the most adequate value (Finlay, 2002; Fenchel and Finlay, 2004). The biological assumption underlying this abrupt threshold is that microscopic organisms (and also macroscopic organisms with microscopic propagules such as fungi or bryophytes) are really different from larger ones (Fontaneto, 2011): they are so small that they can be passively dispersed everywhere, produce resting stages that allow them to survive adverse conditions and persist in any habitat, and can use asexual or parthenogenetic reproduction to quickly increase in number. According to the so-called 'Everything-is-Everywhere' hypothesis (Beijerink, 1913; Baas-Becking, 1934; Finlay, 2002;

Fenchel and Finlay, 2003; de Wit and Bouvier, 2006), these peculiar features would allow many microscopic organisms to attain cosmopolitan distribution. Such cosmopolitanism is quite uncommon in large organisms, implying that the biogeographic principles derived from macrobes are not general, and therefore that new hypotheses and theory shall be developed for microbes (Jenkins *et al.*, 2011). However, the hypothesis that for microscopic organisms everything is everywhere ('ubiquity hypothesis') is considered too simplistic in the current scientific discussion, in opposition to the 'moderate endemicity model' proposed by Foissner (1999, 2006), which suggests that many microscopic organisms have indeed restricted distributions, mostly in connection to their ecological traits other than size (e.g. Faurby and Funch, 2010; Foissner, 2011).

Current challenges to microbial biogeography

The basic units for any biogeographical analysis are records of the presence of species (or any other biological entity) and species lists (see Hortal, 2008). This poses two fundamental challenges to the development of microbial biogeography. On one hand, identifying microbial 'species' and/or defining meaningful units of diversity. On the other, linking spatial patterns with processes based on recorded presences.

Taxonomy

Most studies supporting the ubiquity hypothesis use morphological characteristics to identify the taxonomic units under consideration (Fenchel and Finlay, 2006). However, finding reliable morphological taxonomic characters for most microscopic organisms is notoriously difficult, so the use of morphological features may result in lumping together biological entities with distinct ecological and biogeographical attributes. This constitutes a fundamental problem, as units of diversity need to be unambiguously defined to map their distribution (Bass and Boenigk, 2011; Lacap *et al.*, 2011). Thus, many species considered cosmopolitan may in fact represent complexes of cryptic species with restricted distributions. There are ways to solve this problem, and new tools help providing reliable estimates of diversity. Molecular taxonomy helps morphological taxonomy to reveal the existence of species complexes, and can potentially identify the correct units of diversity (that is, distinct evolutionary entities) within them (Pons *et al.*, 2006; Burns *et al.*, 2008; Barraclough *et al.*, 2009). Additionally, environmental sequencing (by either cloning PCR products or ultrasequencing) is providing distributional data of many taxa otherwise unrecorded or undistinguished (e.g. Robeson *et al.*, 2009; Creer *et al.*, 2010).

Spatial patterns

For microscopic organisms, it will be difficult to disentangle the contribution of historical versus ecological biogeography (Bass and Boenigk, 2011). Both large and small organisms present small discrepancies between their potential and realized distributions (that is, all the places where they could live in opposition to all the places where they actually live). However, they do it in a radically different way. The realized distribution of macrobe species usually constitutes a subset of their potential distribution, because most of them do not occupy all the places suitable for them (i.e. their distributions are not in equilibrium with the environment; see Araújo and Pearson, 2005). Absence from suitable places could be caused by a number of factors, such as historical effects, limitations to dispersal, or the presence of competitors (see Soberón, 2007, 2010; Soberón and Nakamura, 2009; Lobo *et al.*, 2010). In contrast, the realized distribution of microscopic species may be larger than its potential distribution, because they are often present in areas that are not ecologically optimal for them thanks to their dispersal abilities and their capacity to survive as resting stages (Ganter, 2011; Marchant *et al.*, 2011). This rarely happens in macroscopic organisms, and when it does it happens in areas placed nearby suitable sites, as a result of metapopulational processes (see Soberón and Nakamura, 2009). It follows that disentangling the effects of historical and ecological processes on the geographic distributions of microbes may prove to be an extremely difficult task. Moreover, it highlights the role of the mode of dispersal in creating fundamental differences between the biogeographies of small and large organisms, being also one of the main difficulties in supporting or falsifying the ubiquity hypothesis.

To further complicate the scenario, almost all studies and reviews on the biogeography of microscopic organisms report both evidence of large distributions and of endemic, restricted distributions. The same is true for phylogeographic analyses using DNA sequences to investigate the spatial patterns of distribution, which provide evidence of both long-distance gene flow and restricted gene flow (see reviews for different groups in Artois *et al.*, 2011; Guil, 2011; Medina *et al.*, 2011; Werth, 2011).

In addition to this, the little knowledge on the spatial patterns of variation in microbial communities is scattered among different types of analyses and measurements, such as distance–decay relationships, taxa–area relationships or local:global taxa richness. And these sources of information often offer mixed, when not contradictory, information about the similarities and differences between microbes and macrobes. The heterogeneity in the sources of evidence and other potential confounding effects such as the spatial scale of the analysis limit the explanatory capacity of the works that try to infer the processes driving the spatial patterns of microorganisms (De Meester, 2011; Hortal, 2011; Jenkins *et al.*, 2011).

Empirical evidence

Microscopic organisms encompass almost all major groups of living organisms, as they are defined, for biogeographical purposes, as being smaller than 2 mm. Thus, prokaryotes and unicellular eukaryotes fall into this group, together with microscopic animals and fungi. Moreover, other groups of organisms, larger in size, have microscopic dispersing stages and, for biogeographical purposes, can be considered in this review, like fungi, lichens, mosses and ferns (Fontaneto, 2011).

Prokaryotes

Some recent studies on prokaryotes have attempted to address spatial patterns in bacterial and archaeal taxa with respect to spatial scales, environmental factors and temporal scales (Green and Bohannan, 2006; Hughes-Martiny *et al.*, 2006; Prosser *et al.*, 2007; Lacap *et al.*, 2011). The existence of taxon–area relationships, where species richness increases with the amount of area sampled, has been found for tree hole bacteria and salt marsh bacteria (Horner-Devine *et al.*, 2004; Bell *et al.*, 2005). The existence of a distance-decay pattern for soil prokaryotes has been demonstrated on small scales (metres) (Franklin and Mills, 2003), and on larger scales (Cho and Tiedje, 2000; Reche *et al.*, 2005). Other studies have concluded that the influence of environmental heterogeneity was more important than geographic distance as a driving force shaping community composition (Lacap *et al.*, 2011).

There are even evidences of 'ancient endemism' in prokaryotes (Takacs-Vesbach *et al.*, 2008). Thus, even if diversity in prokaryotes is underestimated and range sizes are generally wider than in larger organisms, biogeographical patterns seem to exist, in contrast to the ubiquity hypothesis.

Unicellular eukaryotes

The main proponents of the ubiquity hypothesis in protists have been Finlay, Fenchel and colleagues (e.g. Finlay *et al.*, 1996; Finlay, 2002; Fenchel and Finlay, 2004; Finlay and Fenchel, 2004). The original formulation of the ubiquity hypothesis for protists uses morphology-based definitions of the taxonomic units (Fenchel and Finlay, 2006). Apart from the problems of such approach discussed above, many protists of different groups are not globally distributed even at the level of morphospecies (Bass and Boenigk, 2011); these groups include unicellular green algae (Coesel and Krienitz, 2008), planktonic foraminifera (Darling and Wade, 2008), testate amoebae (Smith and Wilkinson, 2007; Heger *et al.*, 2011), ciliates (Foissner *et al.*, 2003; Stoeck *et al.*, 2007), diatoms (Vanormelingen *et al.*, 2008), and chrysophytes (Kristiansen, 2008).

Thus, unicellular eukaryotes seem to exhibit a combination of cosmopolitan and restricted distributions, similar to the patterns found in prokaryotes. Even using only a morphological approach in species identification, in his review of the subject Foissner (2011) suggested that the most important factors describing biogeography in protists are the resting cysts and the geological history: that is, a combination of spatial and biological features limiting dispersal capabilities. Further, based on their review of evidence coming from molecular tools, Bass and Boenigk (2011) suggested that 'the most sensible view is that protist distribution is not fundamentally different to that of other organisms – the apparent differences being quantitative rather than qualitative'.

Multicellular eukaryotes

Almost all groups of multicellular microscopic organisms analysed so far show the same trend of prokaryotes and protists, with evidence of both widespread and restricted distributions, most of the time related to biological properties other than size (Guil, 2001; Artois *et al.*, 2011). The few groups analysed at the community level, such as tardigrades and rotifers, also show responses to environmental gradients similar to those of macrobes (Fontaneto and Ricci, 2006; Guil *et al.*, 2009a; Obertegger *et al.*, 2010).

An even stronger biogeographical pattern is present in larger organisms with microscopic dispersing stages, which should in principle allow global distribution, such as fungi (Geml, 2011), lichens (Werth, 2011), mosses (Medina *et al.*, 2011) and ferns (Schaefer, 2011). Thus it is also true for multicellular organisms that being microbial in size or with a microscopic dispersing stage is not the only requisite for cosmopolitan distribution.

Consequences and theoretical implications

A general theory of microbial biogeography is yet lacking. However, despite scarce and sparse, the information on the geographical responses of microscopic organisms may be just enough to develop a theoretical framework that will lay the foundations of such theory. Some efforts have been made in that direction (see also Hughes-Martiny *et al.*, 2006; Green and Bohannan, 2006; Telford *et al.*, 2006; Green *et al.*, 2008), but further integrative work is needed to solve some key questions. Perhaps the most important of these questions is whether everything small is everywhere, because it informs on the spatial dynamics of microbes across scales. It now seems clear that the large dispersal potential of microbes does not necessarily result in high rates of *effective dispersal* (i.e. successful dispersal events; see Weisse, 2008). Owing to this, many microscopic organisms do not maintain significant levels of gene flow between geographically remote populations (Bohonak and Jenkins, 2003; Foissner, 2006, 2008; Jenkins *et al.*, 2007; Frahm, 2008; Weisse, 2008). This results in the geographically structured genetic differences (i.e. phylogeographic variations) that have been found for many microbial taxa (e.g. Whitaker *et al.*, 2003; Mills *et al.*, 2007; Prosser *et al.*, 2007; Vos and Velicer, 2008).

Another consequence of the combination of the potential for long-distance dispersal events with limited rates of effective dispersal is the particular scaling of the spatial structure of many microbial distributions. According to De Meester (2011), the geographical distributions of free-living terrestrial microbes are determined by the interplay of processes acting at the metacommunity and biogeographical levels. Within a region, the structure of microbial metacommunities is driven by their high dispersal, typically in a gradient between species sorting and mass effects; here, species sorting occurs when dispersal is sufficient to allow the species to colonize all suitable patches but not so high to allow maintaining populations in unsuitable patches, and mass effects when dispersal is so high that it allows maintaining populations in unsuitable patches through source-sink dynamics, reducing the match between the distribution of species and the composition of communities and the local environmental conditions (see Leibold *et al.*, 2004). Between geographically distant regions, monopolization processes will be most important (De Meester, 2011). Monopolization arises as a result of the combination of high dispersal rates with the high population growth rates, short generation times and capacity to produce dormant stage banks that are typical of many microscopic organisms. The combination of these characteristics may often result in strong numerical advantage of first colonizers, which is maintained thanks to the colonization from the bank of dormant individuals and reinforced by local adaptation, providing the resident population with great advantage over new colonizers (De Meester *et al.*, 2002; Urban and De Meester, 2009). This dual process will produce restrictions to species distributions globally (thus allowing the appearance of endemism and phylogeographic structure), while maintaining widespread distributions regionally. Further, it provides an explanation for the differences in the strength of species replacement between patches of different habitats found for some microbial groups (e.g. Fontaneto *et al.*, 2006) as well as for the importance of distance in determining such replacement between environmentally different sites (e.g. Verleyen *et al.*, 2009): the similarity in species composition between different habitats will increase depending on how strong mass effects are for the majority of species within the landscape.

Following De Meester's (2011) framework, another major difference between microbes and macrobes lies in the different *magnitude* of the distance-dependent changes in species composition across similar habitats. The much larger distances needed to find similar degrees of compositional changes result in a completely different scaling between the macroecological relationships observed in microbes and macrobes, despite other processes being similar in both groups. A good example is the relationship between diversity and area, one of the most studied relationships in the biogeography of large organisms. The number of species (or any other diversity unit) typically increases with area; the larger the area, the more species are present. This is true for both discrete patches of habitat that differ in area (i.e. larger ones host more species), and for progressively larger portions of an apparently homogeneous territory. While the former mostly relates to the capacity of a territory of hosting more species (i.e. carrying capacity), the latter is due to variations in the species composition of the communities throughout space (see reviews in Rosenzweig, 1995; Whittaker and Fernández-Palacios, 2007; Lomolino *et al.*, 2010). Strikingly, currently available evidence indicates that while such spatial replacement in the composition of local communities occurs at much larger scales for microbes than for macrobes (Finlay *et al.*, 1998; Green *et al.*, 2004), the strength of taxa–area relationships for discrete habitat islands is similar in both groups (Bell *et al.*, 2005; van der Gast, 2005; see also Green and Bohannan, 2006; Prosser *et al.*, 2007). This indicates that although the composition of microbial communities changes with distance at a lower rate than macrobes, the restrictions to diversity imposed by area may be similar for both groups of organisms (Hortal, 2011).

Other macroecological relationships well-known for macrobes are also relevant to microbes. The diversity of microbial communities varies along environmental gradients; in particular, altitudinal variations in richness have been found at least for rotifers, tardigrades and bryophytes (Fontaneto and Ricci, 2006; Fontaneto *et al.*, 2006; Guil *et al.*, 2009a; Oliveira *et al.*, 2009; Obertegger *et al.*, 2010). These climate-driven diversity variations in microbes may simply be the result of species sorting, due to the differences in niche requirements of each of the species regionally available (see above). In fact, in two of the examples given above many species of both bdelloid rotifers and tardigrades show strong habitat selection (i.e. habitat sorting; Fontaneto and Ricci, 2006; Guil *et al.*, 2009b). This is consistent with a high potential for dispersal and local environmental selection, thereby being more likely to occur when many species are closer to species sorting-like metacommunity dynamics in the framework proposed by De Meester (2011). However, the limited information available, which is scattered between empirical and experimental studies of different groups and types of habitats, impedes the formation of a solid conclusion on the mechanisms driving the dependence of microscopic organism diversity on both area and environmental gradients.

Regardless of the origin of geographical gradients, a straightforward consequence of the spatial variations described above is the existence of geographic differences in the ecosystem functions provided by microbial communities. Different aspects of the diversity of microbial communities, from richness to composition or functional structure, affect ecosystem productivity and functioning (Laakso and Setälä, 1999; Fukami and Morin, 2003; Heemsbergen *et al.*, 2004; Sánchez-Moreno *et al.*, 2008). Given the importance of microscopic organisms for performing and/or maintaining many ecosystem services, the geographic variations in microbial diversity will have important functional consequences (Naeslund and Norberg, 2006; Green *et al.*, 2008). However, like many other aspects of microbial biogeography there is as yet little knowledge about the large-scale impacts of these geographic differences in the ecological functions performed by microscopic organisms.

Outlook

Although the first observations on the geographical distribution of microscopic organisms date from the Twentieth Century, the discipline of microbial biogeography is still in its infancy. The distributions, genetic structure and community variations in space have been comprehensively studied only since the 1990s; in part thanks to the re-opening of the debate on whether everything small is present everywhere. In spite of the limited amount of information available and its sparse character, evidence points clearly to the existence of biogeographical patterns in microscopic organisms. However, these patterns may be markedly different from those observed for macrobes. These differences are not only generated by their different size, but also by the basic ecological differences between both types of organisms. In contrast with the majority of large organisms, microbes have a great potential for dispersal; they are able to produce massive numbers of propagules and to disperse long distances passively in dormant stages. This allows them to reach localities placed much farther apart than any macroscopic organism, and at the same time allows colonizing species to monopolize a locality or region by flooding the available habitats with their propagules, thus impeding other species to establish populations. These particularities result in much lower rates of species replacement with distance compared with those observed for macrobes, as well as in higher rates of establishing populations in environmentally suboptimal habitats. But, at the same time, the same processes produce macroecological responses of diversity to area and environmental gradients that are in essence similar to those of macrobes. Further research is yet needed to determine the generality of these processes and the development of a robust body of theory that allows the establishment of microbial biogeography as a mature research field.

References

Araújo, M.B., and Pearson, R.G. (2005). Equilibrium of species' distributions with climate. Ecography *28*, 693–695.

Artois, T., Fontaneto, D., Hummon, W.D., McInnes, S.J., Todaro, M.A., Sørensen, M.V., and Zullini, A. (2011). Ubiquity of microscopic animals? Evidence from the morphological approach in species identification. In Biogeography of Microscopic Organisms: is Everything Small Everywhere? Fontaneto, D., ed. (Systematics Association and Cambridge University Press, Cambridge, UK), pp. 244–283.

Baas-Becking, L.G.M. (1934). Geobiologie of Inleiding tot de Milieukunde (W.P. van Stockum and Zoon, The Hague, the Netherlands).

Barraclough, T.G., Hughes, M., Ashford-Hodges, N., and Fujisawa, T. (2009). Inferring evolutionarily significant units of bacterial diversity from broad environmental surveys of single-locus data. Biol. Lett. *5*, 425–428.

Bass, D., and Boenigk, J. (2011). Everything is Everywhere: a twenty-first century de-reconstruction with respect to protists. In Biogeography of Microscopic Organisms: is Everything Small Everywhere? Fontaneto, D., ed. (Systematics Association and Cambridge University Press, Cambridge, UK), pp. 88–110.

Beijerinck, M.W. (1913). De infusies en de ontdekking der backteriën. Jaarboek van de Koninklijke Akademie voor Wetenschappen (Müller, Amsterdam, the Netherlands) (reprinted in Verzamelde geschriften van M.W. Beijerinck, vijfde deel, Delft, 1921), pp. 119–140.

Bell, T., Ager, D., Song, J.-I., Newman, J.A., Thompson, I.P., Lilley, A.K., and van der Gast, C.J. (2005). Larger islands house more bacterial taxa. Science *308*, 1884.

Bergmann, C. (1847). Über die verhältnisse der wärmeökonomie der thiere zu ihrer grösse. Göttinger Studien *3*, 595–708.

Bohonak, A.J., and Jenkins, D.G. (2003). Ecological and evolutionary significance of dispersal by freshwater invertebrates. Ecol. Lett. *6*, 783–796.

Bonner, J.T. (2006). Why size matters: from bacteria to blue whales (Princeton University Press, Princeton, NJ).

Brown, J.H., and West, G.B. (2000). Scaling in biology (Oxford University Press, New York).

Brown, J.H., Gillooly, J.F., Allen, A.P., Savage, V.M., and West, G.B. (2004). Toward a metabolic theory of ecology. Ecology *85*, 1771–1789.

Burns, J.M., Janzen, D.H., Hajibabaei, M., Hallwachs, W., and Hebert, P.D.N. (2008). DNA barcodes and cryptic species of skipper butterflies in the genus *Perichares* in Area de Conservacion Guanacaste, Costa Rica. Proc. Natl. Acad. Sci. U.S.A. *105*, 6350–6355.

Calder, W.A., III. (1996). Size, Function and Life History (Dover Publications, Mineola, NY).
Cardillo, M., and Bromhan, L. (2001). Body size and risk of extinction in Australian mammals. Conserv. Biol. *15*, 1435–1440.
Cho, J.-C., and Tideje, J.M. (2000). Biogeography and degree of endemicity of fluorescent *Pseudomonas* strains. Appl. Environ. Microb. *66*, 5448–5456.
Coesel, P.F.M., and Krienitz, L. (2008). Diversity and geographic distribution of desmids and other coccooid green algae. Biodivers. Conserv. *17*, 381–392.
Colinvaux, P. (1978). Why big fierce animals are rare (Princeton University Press, Princeton, NJ).
Creer, S., Fonseca, V.G., Porazinska, D.L., Giblin-Davis, R.M., Sung, W., Power, D.M., Packer, M., Carvalho, G.R., Blaxter, M.L., Lambshead, P.J.D., *et al.* (2010). Ultrasequencing of the meiofaunal biosphere: practice, pitfalls and promises. Mol. Ecol. *19*, 4–20.
Darling, K.F., and Wade, C.M. (2008). The genetic diversity of planktonic foraminifera and the global distribution of ribosomal RNA genotypes. Mar. Micropaleontol. *67*, 216–238.
De Meester, L. (2011). A metacommunity perspective on the phylo- and biogeography of small organisms. In Biogeography of Microscopic Organisms: is Everything Small Everywhere? Fontaneto, D., ed. (Systematics Association and Cambridge University Press, Cambridge, UK), pp. 324–334.
De Meester, L., Gómez, A., Okamura, B., and Schwenk, K. (2002). The Monopolization Hypothesis and the dispersal-gene flow paradox in aquatic organisms. Acta Oecol. *23*, 121–135.
Diniz-Filho, J.A.F., Rodríguez, M.Á., Bini, L.M., Olalla-Tárraga, M.Á., Cardillo, M., Nabout, J.C., Hortal, J., and Hawkins, B.A. (2009). Climate history, human impacts and global body size of Carnivora (Mammalia: Eutheria) at multiple evolutionary scales. J. Biogeogr. *36*, 2222–2236.
Faurby, S., and Funch, P. (2010). Size is not everything: A meta-analysis of geographic variation in microscopic eukaryotes. Global Ecol. Biogeogr. *20*, 475–485.
Fenchel, T., and Finlay, B.J. (2003). Is microbial diversity fundamentally different from biodiversity of larger animals and plants? Eur. J. Protistol. *39*, 486–490.
Fenchel, T., and Finlay, B.J. (2004). The ubiquity of small species: patterns of local and global diversity. BioScience *54*, 777–784.
Fenchel, T., and Finlay, B.J. (2006). The diversity of microbes: resurgence of the phenotype. Philos. T. R. Soc. B *361*, 1965–1973.
Finlay, B.J. (2002). Global dispersal of free-living microbial eukaryote species. Science *296*, 1061–1063.
Finlay, B.J., and Esteban, G.F. (2007). Body size and biogeography. In Body Size: the Structure and Function of Aquatic Ecosystems, Hildrew, A., Raffaelli, D., and Edmonds-Brown, R., eds (Cambridge University Press, Cambridge, UK), pp. 167–185.
Finlay, B.J., and Fenchel, T. (1999). Divergent perspectives on protist species richness. Protist *150*, 229–233.
Finlay, B.J., Corliss, J.O., Esteban, G., and Fenchel, T. (1996). Biodiversity at the microbial level: the number of free-living ciliates in the biosphere. Q. Rev. Biol. *71*, 221–237.
Finlay, B.J., Esteban, G.F., and Fenchel, T. (1998). Protozoan diversity: converging estimates of the global number of free-living ciliate species. Protist *149*, 29–37.
Foissner, W. (1999). Protist diversity: estimates of the near-imponderable. Protist *150*, 363–368.
Foissner, W. (2006). Biogeography and dispersal of micro-organisms: A review emphasizing protists. Acta Protozool. *45*, 111–136.
Foissner, W. (2008). Protist diversity and distribution: some basic considerations. Biodivers. Conserv. *17*, 235–242.
Foissner, W. (2011). Dispersal of protists: the role of cysts and human introductions. In Biogeography of Microscopic Organisms: is Everything Small Everywhere? Fontaneto, D., ed. (Systematics Association and Cambridge University Press, Cambridge, UK), pp. 61–87.
Foissner, W., Strüde-Kypke, M., van der Staay, G.W.M., Moon-van der Staay, S.-Y., and Hackstein, J.H.P. (2003). Endemic ciliates (Protozoa, Ciliophora) from tank bromeliads (Bromeliaceae): a combined morphological, molecular, and ecological study. Eur. J. Protistol. *39*, 365–372.
Fontaneto, D. (ed.) (2011). Biogeography of Microscopic Organisms: is Everything Small Everywhere? (Systematics Association and Cambridge University Press, Cambridge, UK).
Fontaneto, D., and Ricci, C. (2006). Spatial gradients in species diversity of microscopic animals: the case of bdelloid rotifers at high altitude. J. Biogeogr. *33*, 1305–1313.
Fontaneto, D., Ficetola, G.F., Ambrosini, R., and Ricci, C. (2006). Patterns of diversity in microscopic animals: are they comparable to those in protists or in larger animals? Global Ecol. Biogeogr. *15*, 153–162.
Frahm, J.P. (2008). Diversity, dispersal and biogeography of bryophytes (mosses). Biodivers. Conserv. *17*, 277–284.

Franklin, R.B., and Mills, A.L. (2003). Multi-scale variation in spatial heterogeneity for microbial community structure in an eastern Virginia agricultural field. FEMS Microbiol. Ecol. *44*, 335–346.

Fukami, T., and Morin, P.J. (2003). Productivity–biodiversity relationships depend on the history of community assembly. Nature *424*, 423–426.

Ganter, P.F. (2011). Everything is not everywhere: the distribution of cactophilic yeast. In Biogeography of Microscopic Organisms: is Everything Small Everywhere? D. Fontaneto, ed. (Cambridge, UK: Systematics Association and Cambridge University Press), pp. 130–174.

van der Gast, C.J., Lilley, A.K., Ager, D., and Thompson, I.P. (2005). Island size and bacterial diversity in an archipelago of engineering machines. Environ. Microbiol. *7*, 1220–1226.

Geml, J. (2011). Coalescent analyses reveal contrasting patterns of inter-continental gene flow in arctic-alpine and boreal-temperate fungi. In Biogeography of Microscopic Organisms: is Everything Small Everywhere? Fontaneto, D., ed. (Systematics Association and Cambridge University Press, Cambridge, UK), pp. 177–190.

Green, J., and Bohannan, B.J.M. (2006). Spatial scaling of microbial biodiversity. Trends Ecol. Evol. *21*, 501–507.

Green, J.L., Holmes, A.J., Westoby, M., Oliver, I., Briscoe, D., Dangerfield, M., Gillings, M., and Beattie, A.J. (2004). Spatial scaling of microbial eukaryote diversity. Nature *432*, 747–750.

Green, J.L., Bohannan, B.J.M., and Whitaker, R.J. (2008). Microbial biogeography: from taxonomy to traits. Science *320*, 1039–1043.

Guil, N. (2011). Molecular approach to micrometazoans. Are they here, there and everywhere? In Biogeography of Microscopic Organisms: is Everything Small Everywhere? Fontaneto, D., ed. (Systematics Association and Cambridge University Press, Cambridge, UK), pp. 284–306.

Guil, N., Hortal, J., Sánchez-Moreno, S., and Machordom, A. (2009a). Effects of macro and micro-environmental factors on the species richness of terrestrial tardigrade assemblages in an Iberian mountain environment. Landscape Ecol. *24*, 375–390.

Guil, N., Sánchez-Moreno, S., and Machordom, A. (2009b). Local biodiversity patterns in micrometazoans: Are tardigrades everywhere? Syst. Biodivers. *7*, 259–268.

Hawkins, B.A., Diniz-Filho, J.A.F., Bini, L.M., Araújo, M.B., Field, R., Hortal, J., Kerr, J.T., Rahbek, C., Rodríguez, M.Á., and Sanders, N.J. (2007). Metabolic theory and diversity gradients: Where do we go from here? Ecology *88*, 1898–1902.

Heemsbergen, D.A., Berg, M.P., Loreau, M., van Haj, J.R., Faber, J.H., and Verhoef, H.A. (2004). Biodiversity effects on soil processes explained by interspecific functional dissimilarity. Science *306*, 1019–1020.

Heger T.J., Lara, E., and Mitchell, E.A.D. (2011). Arcellinida testate amoebae (Arcellinida: Amoebozoa): model of organisms for assessing microbial biogeography. In Biogeography of Microscopic Organisms: is Everything Small Everywhere? Fontaneto, D., ed. (Systematics Association and Cambridge University Press, Cambridge, UK), pp. 111–129.

Hildrew, A.G., Raffaelli, D.G., and Edmonds-Brown, R. (2007). Body size: the structure and function of aquatic ecosystems (Cambridge University Press, Cambridge, UK).

Horner-Devine, C., Lange, M., Hughes, J.B., and Bohannan, B.J.M. (2004). A taxa–area relationship for bacteria. Nature *152*, 750–753.

Hortal, J. (2008). Uncertainty and the measurement of terrestrial biodiversity gradients. J. Biogeogr. *35*, 1355–1356.

Hortal, J. (2011). Geographical variation in the diversity of microbial communities: research directions and prospects for experimental biogeography. In Biogeography of Microscopic Organisms: is Everything Small Everywhere? Fontaneto, D., ed. (Systematics Association and Cambridge University Press, Cambridge, UK), pp. 335–357.

Hughes-Martiny, J.B., Bohannan, B.J.M., Brown, J.H., Colwell, R.K., Fuhrman, J.A., Green, J.L., Horner-Devine, M.C., Kane, M., Adams-Krumins, J., Kuske, *et al.* (2006). Microbial biogeography: putting micro-organisms on the map. Nat. Rev. Microbiol. *4*, 102–112.

Jenkins, D.G., Brescacin, C.R., Duxbury, C.V., Elliott, J.A., Evans, J.A., Grablow, K.R., Hillegass, M., Lyon, B.N., Metzger, G.A., Olandese, M.L., *et al.* (2007). Does size matter for dispersal distance? Global Ecol. Biogeogr. *16*, 415–425.

Jenkins, D.J., Medley, K.A., and Franklin, R.B. (2011). Microbes as a test of biogeographical principles. In Biogeography of Microscopic Organisms: is Everything Small Everywhere? Fontaneto, D., ed. (Systematics Association and Cambridge University Press, Cambridge, UK), pp. 309–323.

Kristiansen, J. (2008). Dispersal and biogeography of silica-scaled chrysophytes. Biol. Conserv. *17*, 419–26.

Laakso, J., and Setälä, H. (1999). Sensitivity of primary production to changes in the architecture of belowground food webs. Oikos *87*, 57–64.

Lacap, D.C., Lau, M.C.Y., and Pointing, S.B. (2011). Biogeography of prokaryotes. In Biogeography of Microscopic Organisms: is Everything Small Everywhere? Fontaneto, D., ed. (Systematics Association and Cambridge University Press, Cambridge, UK), pp. 35–42.

Leibold, M.A., Holyoak, M., Mouquet, N., Amarasekare, P., Chase, J.M., Hoopes, M.F., Holt, R.D., Shurin, J.B., Law, R., Tilman, D., *et al.* (2004). The metacommunity concept: a framework for multi-scale community ecology. Ecol. Lett. *7*, 601–613.

Lobo, J.M., Jiménez-Valverde, A., and Hortal, J. (2010). The uncertain nature of absences and their importance in species distribution modelling. Ecography *33*, 103–114.

Lomolino, M.V., Riddle, B.R., Whittaker, R.J., and Brown, J.H. (2010). Biogeography. Fourth Edition (Sinauer Associates, Inc., Sunderland, MA).

Marchant, R., Banat, I.M., and Franzetti, A. (2011). Thermophilic bacteria in cool soils: metabolic activity and mechanisms of dispersal. In Biogeography of Microscopic Organisms: is Everything Small Everywhere? Fontaneto, D., ed. (Systematics Association and Cambridge University Press, Cambridge, UK), pp. 43–57.

Martínez del Rio, C. (2008). Metabolic theory or metabolic models? Trends Ecol. Evol. *23*, 256–260.

Medina, N.G., Draper, I., and Lara, F. (2011). Biogeography of mosses and allies: does size matter? In Biogeography of Microscopic Organisms: is Everything Small Everywhere? Fontaneto, D., ed. (Systematics Association and Cambridge University Press, Cambridge, UK), pp. 209–233.

Meiri, S., and Dayan, T. (2003). On the validity of Bergmann's rule. J. Biogeogr. *30*, 331–351.

Mills, S., Lunt, D.H., and Gomez, A. (2007). Global isolation by distance despite strong regional phylogeography in a small metazoan. BMC Evol. Biol. *7*, 225.

Muller-Landau, H.C., Condit, R.S., Chave, J., Thomas, S.C., Bohlman, S.A., Bunyavejchewin, S., Davies, S., Foster, R., Gunatilleke, S., Gunatilleke, N., *et al.* (2006). Testing metabolic ecology theory for allometric scaling of tree size, growth and mortality in tropical forests. Ecol. Lett. *9*, 575–588.

Naeslund, B., and Norberg, J. (2006). Ecosystem consequences of the regional species pool. Oikos *115*, 504–512.

Obertegger, U., Thaler, B., and Flaim, G. (2010). Rotifer species richness along an altitudinal gradient in the Alps. Global Ecol. Biogeogr. *19*, 895–904.

Olalla-Tárraga, M.A., Rodríguez, M.A., and Hawkins, B.A. (2006). Broad-scale patterns of body size in squamate reptiles of Europe and North America. J. Biogeogr. *33*, 781–793.

Oliveira, S.M., ter Steege, H., Cornelissen, J.H.C., and Gradstein, S.R. (2009). Niche assembly of epiphytic bryophyte communities in the Guianas: a regional approach. J. Biogeogr. *36*, 2076–2084.

O'Malley, M.A. (2008). '*Everything is everywhere*: but *the environment selects*': Ubiquitous distribution and ecological determinism in microbial biogeography. Stud. Hist. Philos. Biol. Biomed. Sci. *39*, 314–325.

Peters, R.H. (1986). The ecological implications of body size. Cambridge Studies in Ecology (Cambridge University Press, Cambridge, UK).

Pons, J., Barraclough, T.G., Gomez-Zurita, J., Cardoso, A., Duran, D.P., Hazell, S., Kamoun, S., Sumlin, W.D., and Vogler, A.P. (2006). Sequence-based species delimitation for the DNA taxonomy of undescribed insects. Syst. Biol. *55*, 595–609.

Prosser, J.I., Bohannan, B.J.M., Curtis, T.P., Ellis, R.J., Firestone, M.K., Freckleton, R.P., Green, J.L., Green, L.E., Killham, K., Lennon, J.J., *et al.* (2007). The role of ecological theory in microbial ecology. Nat. Rev. Microbiol. *5*, 384–392.

Reche, I., Pulido-Villena, E., Morales-Bacquero, R., and Casamayor, E.O. (2005). Does ecosystem size determine aquatic bacterial richness? Ecology *86*, 1715–1722.

Robeson, M.S., II, Costello, E.K., Freeman, K.R., Whiting, J., Adams, B., Martin, A.P., and Schmidt, S.K. (2009). Environmental DNA sequencing primers for eutardigrades and bdelloid rotifers. BMC Ecol. *9*, 25.

Rosenzweig, M.L. (1995). Species diversity in space and time (Cambridge University Press, Cambridge, UK).

Sánchez-Moreno, S., Ferris, H., and Guil, N. (2008). Role of tardigrades in the suppressive service of a soil food web. Agr. Ecosyst. Environ. *124*, 187–192.

Schaefer, H. (2011). Dispersal limitation or habitat quality – what shapes the distribution ranges of ferns? In Biogeography of Microscopic Organisms: is Everything Small Everywhere? Fontaneto, D., ed. (Systematics Association and Cambridge University Press, Cambridge, UK), pp. 234–243.

Schmidt-Nielsen, K. (1984). Scaling: why is animal size so important? (Cambridge University Press, Cambridge, UK).

Smith, H.G., and Wilkinson, D.M. (2007). Not all free-living micro-organisms have cosmopolitan distributions – the case of *Nebela* (*Apodera*) *vas* Certes (Protozoa: Amoebozoa: Arcellinida). J. Biogeogr. *34*, 1822–1831.

Soberón, J. (2007). Grinnellian and Eltonian niches and geographic distributions of species. Ecol. Lett. *10*, 1115–1123.

Soberón, J. (2010). Niche and area of distribution modeling: A population ecology perspective. Ecography *33*, 159–167.

Soberón, J., and Nakamura, M. (2009). Niches and distributional areas: Concepts, methods, and assumptions. Proc. Natl. Acad. Sci. U.S.A. 106, 19644–19650.

Stoeck, T., Bruemmer, F., and Foissner, W. (2007). Evidence for local ciliate endemism in an Alpine anoxic lake. Microb. Ecol. *54*, 478–486.

Takacs-Vesbach, C., Mitchell, K., Jackson-Weaver, O., and Reysenbach, A.-L. (2008). Volcanic calderas delineate biogeographic provinces among Yellowstone thermophiles. Environ. Microbiol. *10*, 1681–1689.

Telford, R.J., Vandvik, V., and Birks, H.J.B. (2006). Dispersal limitations matter for microbial morphospecies. Science *312*, 1015.

Urban, M.C., and De Meester, L. (2009). Community monopolization: local adaptation enhances priority effects in an evolving metacommunity. P. Roy. Soc. B – Biol. Sci. *276*, 4129–4138.

Vanormelingen, P., Verleyen, E., and Vyverman, W. (2008). The diversity and distribution of diatoms: from cosmopolitanism to narrow endemism. Biodivers. Conserv. *17*, 393–405.

Verleyen, E., Vyverman, W., Sterken, M., Hodgson, D.A., Wever, A.D., Juggins, S., Vijver, B.V. d., Jones, V.J., Vanormelingen, P., Roberts, D., *et al.* (2009). The importance of dispersal related and local factors in shaping the taxonomic structure of diatom metacommunities. Oikos *118*, 1239–1249.

Vos, M., and Velicer, G.J. (2008). Isolation by distance in the spore-forming soil bacterium *Myxococcus xanthus*. Curr. Biol. *18*, 386–391.

Weisse, T. (2008). Distribution and diversity of aquatic protists: an evolutionary and ecological perspective. Biodivers. Conserv. *17*, 243–259.

Werth, S. (2011). Biogeography and phylogeography of lichen fungi and their photobionts. In Biogeography of Microscopic Organisms: is Everything Small Everywhere? D. Fontaneto, ed. (Cambridge, UK: Systematics Association and Cambridge University Press), pp. 191–208.

Whitaker, R.J., Grogan, D.W., and Taylor, J.W. (2003). Geographic barriers isolate endemic populations of hyperthermophilic archaea. Science *301*, 976–978.

Whittaker, R.J., and Fernández-Palacios, J.M. (2007). Island Biogeography: Ecology, Evolution, and Conservation. Second edition (Oxford University Press, Oxford, UK).

Williams, D.M. (2011). Historical biogeography, microbial endemism and the role of classification: everything is endemic. In Biogeography of Microscopic Organisms: is Everything Small Everywhere? Fontaneto, D., ed. (Systematics Association and Cambridge University Press, Cambridge, UK), pp. 11–31.

de Wit, R., and Bouvier, T. (2006). 'Everything is everywhere, but, the environment selects'; what did Baas Becking and Beijerinck really say? Environ. Microbiol. *8*, 755–758.

Index

A

D

E

F

G

I

J

K

L

M

Q

R

S